Chris E. Petersen • Lynda A. Randa • Nancy A. Kett

Laboratory and Field Inquiries in Biology

Second Edition

Cover Art: Courtesy of Corbis, PhotoDisc/Getty Images.
Interior images by Nancy A. Kett.

Copyright © 2015 by Pearson Learning Solutions

All rights reserved.

Permission in writing must be obtained from the publisher before any part of this work may be reproduced or transmitted in any form or by any means, electronic or mechanical, including photocopying and recording, or by any information storage or retrieval system.

All trademarks, service marks, registered trademarks, and registered service marks are the property of their respective owners and are used herein for identification purposes only.

Pearson Learning Solutions, 501 Boylston Street, Suite 900, Boston, MA 02116
A Pearson Education Company
www.pearsoned.com

Printed in the United States of America

000200010271967125

EEB

ISBN 10: 1-323-10539-5
ISBN 13: 978-1-323-10539-9

Contents

1	An Introduction to Biometry	1
2	Writing a Scientific Paper	9
3	Classification and Structure of the Plants	13
4	Classification of the Animals	31
5	Dissection of the Frog	57
6	Dissection of the Fetal Pig	67
7	Animal Tissues	81
8	The Relationship of Environmental Variations to Stomata Density	93
9	Analyzing the Structure of Plant Communities	99
10	Allelopathy	111
11	Measuring the Distributional Patterns of the Tall Goldenrod and Gall-Making Host	115
12	A Field Study of the Reproductive Behavior of Red-Winged Blackbirds	119
13	A Field Investigation on the Seed Foraging Behavior of Vertebrates in the Natural Environment	129
14	Pollination and the Distribution of Plants	137
15	A Laboratory Investigation of Population Growth	141
16	Life Tables	151
Appendix 1	Critical t Table	155
Appendix 2	Critical F Table	156
Appendix 3	Critical U Table for Mann-Whitney Test	157
Appendix 4	Critical Chi-square Table	158

1 An Introduction to Biometry

Purpose

To be able to understand the use of statistics in data summarization and interpretation.

To develop skills in quantitative analysis.

Materials

Scientific calculator

Introduction

Biometry involves the use of statistical analysis in the summarization and interpretation of outcomes from biological experimentation. In this lab, you will be introduced to basic techniques in biometry that will come in handy when analyzing experimental outcomes. Consult textbooks specific to biometry or statistics for more thorough coverage. Including one or more statistics courses in your academic curriculum may be required and should prove essential in understanding the significance of research conducted by others or yourself.

Basic Terminology

A "statistical" population is defined as the collection of items that are to be sampled. The sample should be of sufficient size so that conclusions reached from the sample apply to the parent population from which the sample was taken. For example, say you were interested in the age structure of Northern Pike in a lake as a means to investigate if the fish population was growing, declining, or stable. An age structure provides a breakdown of how many individuals are not yet reproducing (juveniles), are reproducing (reproductives), and no longer reproducing (post-reproductives). A large number of juveniles relative to the older age groups suggests a period of population growth. You may not be able to sample all of the Northern Pike in the lake. However, due to physical, time, and/or economic constraints, analysis of 100 pike could provide sufficient information of the age structure of Northern Pike population in the lake. These 100 individuals would compose the sample. Age might be estimated by measuring length. The mean length of pike can then be computed by,

$$\bar{x} = \frac{\Sigma x_i}{n},$$

where Σ is a symbol for summation (e.g., the length of all of the 100 pike are summed or added together), x_i is a measurement from item "i" (e.g., there

Table 1.1
Length of the Two Shells from Each of Six Clams
Data summarization and results of one-tailed Student t-test comparing shell length per clam are given.

Clam	Length (cm) Shell 1	Shell 2	Difference in Shell Length |Shell 1 − Shell 2|	x_i^2
1	4.1	4.2	0.1	0.01
2	4.0	4.0	0	0
3	3.8	3.7	0.1	0.01
4	3.8	3.8	0	0
5	4.0	4.0	0	0
6	4.1	4.1	0	0
			$\Sigma x_i = 0.2$	$\Sigma(x_i^2) = 0.02$

H_o: Mean difference in shell length = 0.

are 100 items "i" or pike), and n is the sample size (e.g., 100 pike). A measure of how much variance there is among measurements may be quantified by sample standard deviation (SD) which, if subtracted from or added to the mean, provide a measure of the degree to which individuals in the sample differ from the sample mean. The sample standard deviation can be computed by the formula,

$$SD = \sqrt{\frac{\Sigma x_i^2 - (\Sigma x_i)^2/n}{n-1}}.$$

Take a look at Table 1.1 for an example of how the measurements are computed.

Statistical testing of data implies testing of a null hypothesis (H_o) or hypothesis of no difference, effect, or relationship. If you were testing how elevated levels of carbon dioxide affect plant growth, H_o would state that elevated levels of carbon dioxide do not increase plant growth.

Statistical tests include "parametric" tests and "nonparametric" tests. Parametric tests make the assumption that the two populations being compared are normally distributed and show the same dispersion of data, i.e., they show bell-shaped curves of similar amplitude, slope, and spread. Figure 1.1 illustrates what a normal distribution may look like. Parametric tests cannot be used if the populations of items to be compared lack a normal distribution of data and/or fail to show similar variation (homogeneity of

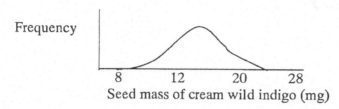

Figure 1.1
Example of how a normal distribution might look.

variance). The researcher can then opt for nonparametric tests which do not require these assumptions, but often at the expense of losing some information concerning the true distribution of the data.

Part A–Parametric Testing of Hypotheses
The Student T-Test

1. One-sample hypothesis testing

   The one-sample t-test is applicable when one of the population means is hypothetical and is specified by the null hypothesis. In order to illustrate computations, consider an experiment which tests if clams are truly bilateral. Assume the difference in lengths from the anterior-most to posterior-most edge of the two shells provides the measurement for comparison. Measurements from six clams and statistical procedures are provided in Table 1.1. The null hypothesis (H_o) would be that there is no difference in shell lengths of a clam where the mean difference is 0. While there is no interest in the particular study of which shell is longer, the only other "alternative" hypothesis, H_A, is that the mean difference should be greater than 0. Hence, the t-test is said to be a "one-tailed" test based on this alternative hypothesis, where there is one alternative to the null hypothesis, i.e., a mean difference occurring. A two-tailed test would result if the H_A stipulates that the mean difference could be less or greater than 0 (where there are "two" sides to H_A), but this is not the case here as the length of a particular shell is not being investigated.

   The equation for the t-test is, $t_o = |(\bar{x} - \mu)/(SD/\sqrt{n})|$,

   where μ is the mean of the parent population and is specified by the null hypothesis, and t_o is the observed t value. The t_o is then compared to a critical t value, t_c, to test the validity of the null hypothesis. The t_c is found from Appendix 1. To find the t_c for the one-tailed t-test concerning the clams, first determine the degrees of freedom as $n - 1$, or the sample size minus 1. Use the coordinates of the 0.05 probability and degrees of freedom to find t_c. In the clam study, the t_c for 5 degrees of freedom is 2.105. The evaluation of the null hypothesis is as follows.

$$\bar{x} = 0.2/6 = 0.033$$

$$SD = \sqrt{\frac{0.02 - (0.2)^2/6}{6-1}} = 0.052.$$

$$t_o = |\,(0.033 - 0)/(0.0052/\sqrt{6})\,| = 1.554$$

$$t_c = 2.015$$

The t_o value of 1.554 is less than the t_c value of 2.105. Because $t_o < t_c$, H_o cannot be rejected. Shell lengths per clam were not significantly different, supporting the null hypothesis that clams are bilaterally symmetrical.

2. Two-sample hypothesis testing

    The Student t-test can be used to test for differences between two sample means assuming that the variances between the two samples are normally distributed and have the same degree of variability. Consider an experiment which examines how temperature affects the rate of fermentation by a set quantity of yeast (Table 1.2). Fermentation rates are measured at 20°C and 37°C. The null hypothesis (H_o) is that temperature has no effect on the mean rate of fermentation. The hypothesis (H_A) alternative to H_o is that mean fermentation rate at 20°C $\neq$ the mean fermentation rate at 37°C. The t-test is thus two-sided as H_A states that the mean rate at 20°C could be less than or greater than the mean rate at 37°C.

    To use the two-sample t-test, homogeneity of variance must be safely assumed for the treatments that are to be compared. In other words and with respect to the example concerning fermentation rates, the variances in fermentation rates under the two temperature regimes should be of the same degree. The "F-test" is designed to examine the question of homogeneity of variance where the sample variance is computed by squaring the standard deviation. Divide the larger of the two variances by the lesser to calculate the F_o value,

    $$F_o = (SD_1^2/SD_2^2).$$

    For the fermentation experiment, $F_o = 0.397^2/0.240^2 = 2.736$. The F_o is then compared to the critical F value, F_c, found in Appendix 2. To locate F_c for the fermentation experiment, determine the degrees of freedom (df) for each sample as $(n - 1)$, where n is the sample size. Hence, the df for each of the two fermentation samples is $(10 - 1)$ or 9. Intersect the column and row using 9 for each coordinate to locate F_c. The F_c for the fermentation experiment is 3.444. If $F_o \leq F_c$, then homogeneity of variance is safely assumed and you may proceed with the t-test. If not, the t-test is not applicable for use. As the $F_o < F_c$ for the fermentation experiment, proceed to use the t-test.

    The two-sample t-test requires computation of two values, the pooled standard error (SE_p) of the difference in the pooled data sets and t_o. SE_p is computed as,

    $$SE_p = \sqrt{\frac{[(n_1-1)SD_1^2 + (n_2-1)SD_2^2][1/n_1 + 1/n_2]}{(n_1 + n_2 - 2)}}$$

    Divide this value into the difference between means to determine t_o as follows,

    $$t_o = |\,(x_1 - x_2)/SE_p\,|.$$

    Locate t_c in Appendix 1 using $df = (n_1 + n_2 - 2)$. If $t_o \geq t_c$, the H_o can be safely rejected. If $t_o < t_c$, H_o cannot be safely rejected. In the fermentation experiment, t_o is greater than t_c, leading to the statistical conclusion that the sample mean fermentation rates are significantly different between temperature treatments.

Table 1.2
CO_2 Evolution (ml/30 seconds) from a yeast culture according to temperature. CO_2 provided an indicator of fermentation rate. The mean fermentation rates were tested using a two-sample t-test to examine if the means are significantly different.

Temperature	
20°	**37°**
1.90	3.53
1.82	3.60
1.96	2.79
2.10	2.78
2.00	2.47
2.01	2.85
1.33	2.93
1.51	2.50
1.93	2.53
1.86	2.99

Sample means ± sample standard deviations
20°C: 1.842 ± 0.240
37°C: 2.897 ± 0.395

$H_o: x_{20} = x_{37}$

Temperature	20°C	37°C
Sample mean	$\frac{18.42}{10} = 1.842$	$\frac{28.97}{10} = 2.897$
Sample standard deviation	$\sqrt{\frac{34.45 - (18.42^2/10)}{10-1}} = 0.240$	$\sqrt{\frac{85.33 - (28.97^2/10)}{10-1}} = 0.395$

$F_o = 0.395^2/0.240^2 = 2.709$. As $F_c = 3.444$ and $F_o < F_c$, can proceed with two-tailed t-test.

Standard error of difference = $\sqrt{\frac{[(10-1)0.240^2 + (10-1)0.395^2] \times [1/10 + 1/10]}{(10+10-2)}} = 0.146$

$t_o = |(1.842 - 2.897)|/0.146 = 7.226$

$t_c = 2.101$ for 2-tailed t-test and $df = (10 + 10 - 2) = 18$

Since $t_o \geq t_c$, H_o can be safely rejected. Conclusion: there is a significant difference in fermentation rates between the two temperature treatments ($P < 0.05; t = 7.226; df = 18$).

Part B–Non-Parametric Testing of Hypotheses

1. Mann-Whitney test for testing median ranks between two samples

    This test provides an alternative to the t-test and is a recommended test when homogeneity of variance cannot be safely assumed as concluded by the F-test. The Mann-Whitney test uses rank measurements of values, where values are ranked from highest to lowest or lowest to highest (choose which direction you want to proceed). The null hypothesis is based on comparison of median ranks and not the actual means. If values are equal, the values are assigned the average rank that would have been assigned to them if they were not equal. Refer to Table 1.3 for an example of computations. The example uses the data from the fermentation experiment discussed in the prior section. Here, the difference in median ranks is to be tested. In Table 1.3, the data values are ranked from lowest to highest. Twenty values should result in a top rank of 20.

Note in Table 1.3 where the data values are equal. The first occurs where the values are 2.01. These values are assigned the median rank they cover, or 9.5 (median rank of 9 and 10). Hence, ranks 9 and 10 are assigned this median rank. The next higher rate after 2.01 is 2.10. Again, equal values occur. The two values are given the median rank of ranks that would have been assigned to them or 11.5 (median rank of ranks 11 and 12). The next highest fermentation rate is 2.17. There is only one reading of 2.17 and this reading is assigned the next higher rank of 13.

The Mann-Whitney statistics, U for each data set, are calculated according to which ranks were summed from one of the data sets. In the fermentation experiment, the 20°C treatment provided this summation as the (rank) numbers looked smaller and easier to add. U_1, with reference to which of the temperature treatments provided the summed ranks is computed as,

Table 1.3
CO_2 Evolution (ml/30 seconds) from a Yeast Culture According to Temperature
Parentheses enclose ranks for Mann-Whitney testing. Ranking is done from the lowest to highest rate of CO_2 evolution.

Temperature				
20°C	Rank	37°C	Rank	
1.90	5	2.53	15	
1.82	3	2.10	11.5	Sample means ± sample standard deviations
1.96	7	2.79	20	20°C: 1.842 ± 0.240
2.10	11.5	2.78	16	37°C: 2.606 ± 0.397
2.00	8	2.47	14	
2.01	9.5	2.85	17	
1.33	1	2.93	18	
1.51	2	2.01	9.5	
1.93	6	2.17	13	
1.86	4	2.99	19	
	$R = 57$			

Ranks for the 20° C treatment were added just because the numbers looked smaller and easier to add.

H_o: Median rank$_{20°}$ = Median rank$_{37°}$

$U_{20} = [(10 \cdot 10) + (10(10 + 1)/2] - 57 = 98$ and

$U_{37} = (10 \cdot 10) - 98 = 2.$

$U_c = 23$

Since one of the calculated U statistics, i.e., U_{20}, is less than U_c, H_o can be safely rejected. The median ranks of fermentation rates at the two temperatures are significantly different ($P < 0.05; U = 2$).

$$U_1 = (n_1 \cdot n_2) + [n_1(n_1 + 1)/2] - R_1.$$

The second U for the second data set is computed as,

$$U_2 = (n_1 \cdot n_2) - U_1.$$

These two observed U's are compared to a critical U_c taken from Appendix 3. Locate n_1 and n_2 and intersect the column and row which they head to find the U_c. If either U_1 or U_2 is $< U_c$, then H_o can be safely rejected.

In the fermentation experiment, U_2 (U_{37}) was less than U_c. The conclusion was that the median rank fermentation rates were significantly different, a similar conclusion as given by the t-test.

2. Chi-square goodness-of-fit test

This test is applicable when examining if data arranged in a frequency distribution conform to a certain theoretical distribution as determined by the null hypothesis (H_o). The computation of the chi-square statistic is,

$$X_o^2 = \Sigma(O - E)^2/E,$$

Where O is the observed or measured frequency and E is the expected frequency based on the H_o. The X_o^2 is compared to a critical X_c^2 (**Appendix 4**) at a 0.05 probability level determined from a degrees of freedom (df) of $n-1$ where n is the number of classes used in the frequency distribution. If $X_o^2 \geq X_c^2$, then H_o can be rejected as being true. If $X_o^2 < X_c^2$, then H_o cannot be rejected.

Table 1.4 provides an example of computations with an experiment that tests Mendelian assumption about inheritance of an autosomal recessive trait of dwarfism in corn. T is the symbol for the tall allele, while t references the dwarfism allele.

When $df = 1$, the X_o^2 needs adjustment to improve the statistical conclusion. Compute the "continuity-corrected" X_o^2 as,

$$X_o^2 = \Sigma[(|O - E| - 0.5)^2/E].$$

For the genetic cross given in Table 1.4, this correction does not affect the statistical conclusion, but it could prove important in other experiments.

Table 1.4
Chi-square Goodness-of-Fit Testing of Mendelian Assumptions of the Pattern of Inheritance for Dwarfism, an Autosomal Recessive Trait in Corn
The parental cross was $Tt \times Tt$ where T represented the dominant allele of "tall" and t represented the recessive trait of dwarfism.

Out of 200 corn progeny from the genetic cross, 154 were tall and 46 were dwarfs.				
H_o: 75% or 150 of the plant progeny should be tall and 25% or 50 of the plant progeny should be dwarfs.				
Class Category	Observed (O)	Expected (E)	$(O-E)^2/E$	With Yates Correction
Tall corn	154	150	0.107	0.082
Dwarf corn	46	50	0.320	0.245
			$X_o^2 = 0.427$	$X_o^2 = 0.327$
$X_c^2 = 3.841$ for $df = 2 - 1$. As $X_o^2 < X_c^2$, H_o cannot be rejected. The X_o^2 computed using the Yates correction is (0.082 + 0.245) or 0.327. As $X_o^2 < X_c^2$, H_o cannot be rejected.				

Suggested Activities

Become more acquainted with the statistical techniques through application using experimental data.

2 Writing a Scientific Paper

Purpose

To be able to understand how a scientific paper is written and, if instructed, gain experience in writing a scientific paper.

Introduction

The writing of scientific papers is a means for a researcher to communicate findings to the scientific community. These papers typically receive peer review before they are accepted for publication. Each paper is reviewed for scientific validity and also other criteria such as conciseness, proper format, and grammar. The paper should be as concise as possible, but without a loss of necessary information. It should be written in a clear and, hopefully, interesting format that follows specification of the journal in which the author(s) hope to have the paper published. Many papers are rejected for publication if one or more of the above criteria are not met.

The Structure of the Scientific Paper

The eight sections of the scientific paper are the TITLE, ABSTRACT, INTRODUCTION, METHODS, RESULTS, DISCUSSION, LITERATURE CITED, and TABLES/FIGURES.

Title

The title is basically a phrase that explains the work in a limited number of words (as limited by the journal). Field studies often include references to time of year and locality.

Abstract

The abstract is a paper of its own which provides the objective(s) of the study, overview of methodology, results, and conclusions as relating to the objective(s). If you feel that the abstract is repetitious of content of the remaining paper, it should be. Potential readers of a paper often read through the abstract in deciding whether to continue reading the remaining paper. Any reference to methods and results should be in the past tense because the experiment has been completed. This rule applies to the remaining sections of the paper as well.

Introduction

The introduction includes a review of the literature (typically books and peer-reviewed articles from journals) to provide the reader with background information pertinent to understanding your topic. The introduction usually ends with the questions, objectives, or hypotheses addressed by your study. Literature citations have a particular format as specified by the journal in which the paper is intended to be published. Your professor will specify a particular peer-reviewed journal that you can inspect for the format of the citations. Be careful in noting differences in the format of book versus journal citations.

Methods

The methods provide explanation of how the experiment was done. Do not provide a list of materials used, but do identify these materials as you explain methodology. A description of the study site for field studies is often given at the beginning of the methods. Statistical treatment of the data often ends the methods.

Results

The results provide the findings of the study in a summarized format. Raw data are usually not provided. Tables and/or figures are created, if needed, to display the data in a visually understandable format. Tables include charts while figures include diagrams, graphs, pictures, and maps. Place any tables and figures at the very end of your paper, tables being before figures. The editor normally decides where to place the tables and figures into the text based on the format and font of the journal. Do cite the tables and figures in the written part of the results and identify in your writing what the reader is expected to understand from these visual displays. Assume that the reader is semi-ignorant as to what are the take-home messages of the data.

Discussion

Discuss results as to their importance in addressing questions, objectives, or hypotheses of your paper. The emphasis of the discussion is your findings. You may use information from the literature to help support your discussion, but do not get side-stepped in creating another review of the literature. Speculate somewhat as to the significance of your findings, but keep the speculation relatively short. Remember that your data are as you measured them. If you were honest with your methodology, including description of the methods, you should not need to criticize your efforts. Why would a reader bother to remember anything from your paper should you criticize your results? You can acknowledge limits of conclusions and provide direction of what the next step of the study should be that researchers could pursue in the future.

Literature Cited

This section includes the full citations of all references cited in the paper. Be sure to follow the format of a refereed journal specified by your professor. Formats found in scientific journals typically are different than those given in any style manuals.

Tables and Figures

These are at the back of your text. Make sure that you cite any table(s) and/or figure(s) in your text to alert your reader and give a statement of what is to be understood by the table(s) and/or figure(s).

Suggested Activities

Read and summarize an article in a scientific journal in an effort to see how the scientific paper is put together. Write a paper on an experiment done in class. Have the paper reviewed and make the suggested revisions in a rewrite of the paper.

3 Classification and Structure of the Plants

Domain Eucarya—Kingdom Plantae

The first plants date back to 540 million years ago (mya) in the fossil record. Although small, these plants had an advantage in height over algae, giving them greater access to light. Competition for light selected for even taller plants which, in turn, selected for tissues capable of supporting a taller individual. By 440 mya, vascular plants had evolved with tissues specialized in conducting fluids through the plant. Around 380 mya, seed-producing plants appeared. Seeds encase young sporophyte embryos and have unique features that facilitate dispersal and enable persistence through periods of poor environmental conditions. The dominant plants of today, the flowering plants, or angiosperms, first appeared when dinosaurs roamed the landscape some 140 mya. These plants diversified quickly, paralleling the rapid diversification of insects, their major pollinators and above-ground consumers. Of the estimated 350,000 species of plants today, 300,000 are angiosperms.

In the following lab, the classification of the plants is introduced. You will be expected to identify the plant phyla and their characteristic features, that is, if they are vascular, the dominant or most common life stage seen according to phylum, if they are seed producers, and if they flower. Answer the questions on the data sheets.

A. The Nonvascular Plants

The Phyla Bryophyta, Hepatophyta, and Anthocerophyta lack the vascular tissues, i.e., xylem and phloem, of higher plants. In the vascular plants, xylem functions in water and mineral transport, while the phloem conducts photosynthates (products of photosynthesis). Many mosses, however, have central cells called hydroids that conduct water and leptoids that conduct photosynthates. Non-vascular plants also lack lignin in their cell walls. Lignin is a polymer that adds strength to the cell walls of vascular plants. The dominant life stage (= most commonly seen stage) in alternation of generations among the nonvascular plants is the gametophyte that nutritionally supports the sporophyte.

Phylum Bryophyta—mosses

The bryophytes are believed to be the link between the green algae and the vascular plants. Like both green algae and higher plants, the bryophytes have a photosynthetic pigment complement of chlorophyll a, chlorophyll b, and carotenoids. Despite being small, mosses have significant economic and ecological value. Peat is derived from accumulated deposits of partially decayed *Sphagnum* (peat moss). It has served as a source of fuel throughout Eurasia, as a soil conditioner for gardens, and even has been used for its antimicrobial action. Well-preserved human remains have even been found in peat! Although small, mosses contribute significantly to the total productivity of certain ecosystems, including temperate rainforests and tundra.

Mosses inhabit moist areas that include the shady sides of tree trunks, but also extreme environments of the tundra, alpine regions, and deserts. Alternation of generations of the bryophytes uniquely includes a filamentous life stage called the protonema.

Observe the living and preserved specimens of mosses. Be able to distinguish the gametophyte and sporophyte stages.

At 100× magnification, observe the moss protonema slide—note the resemblance to the algae.

Phylum Hepatophyta—liverworts

The common name, liverworts, is based on a superficial similarity of the gametophyte to the lobes of the human liver. This lobed and flattened body of the liverworts is called a thallus. Another distinct feature of the liverworts is the embedded archegonium. Like many mosses, liverworts are commonly found in or near water.

Examine living and preserved specimens.

Examine slides of *Marchantia* and locate the large round antheridia and, on a second slide, the archegonia tucked beneath the cap of the archegoniophore.

Phylum Anthocerophyta—hornworts

The hornworts are so named because of the horn-like shape of the sporophytes. The hornworts are unlike the liverworts in that the archegonia are embedded in the gametophytic tissue instead of being borne on stalks, and growth is indeterminate.

Examine cross section (c.s.) and longitudinal (l.s.) slides of *Anthoceros*. In both slides of the sporophyte, the columella, which is centrally located, is surrounded by spores.

B. The Seedless Vascular Plants

The seedless vascular plants comprise the ferns and their relatives. Many are extinct. Preserved as fossils, these plants give clues to the evolution of the vascular plants.

Phylum Lycophyta—club mosses, spike mosses

The Lycophyta are the most ancient group of vascular plants, dating back over 350 million years. Early lycophytes included tree-sized specimens that towered over 40m in height and had diameters greater than 2m. Today, they are represented by small herbaceous survivors. Lycophytes are commonly called club mosses in reference to the club-shaped strobili on sporophytes of many species.

Observe the specimens of club and spike mosses. Note the biramous branching of stems and absence of vascular tissues in roots and leaves.

Phylum Monilophyta—whisk ferns, horsetails, and ferns

Like their ancestors, monilophytes continue to occupy shaded regions of tropical and temperate forests, as well as marsh and moist open areas. Some have evolved to tolerate arid climates. Present-day monilophytes are mostly small, although some ferns may exceed several meters in height. During the Carboniferous period (299–359 mya) lycophytes and monilophytes composed the dominant flora of forests with some monilophytes (i.e., horsetails) exceeding 15m in height.

Observe the whisk fern specimen. The dominant organ of the whisk fern is the stem. The stem shows dichotomous branching and has scale-like photosynthetic structures that lack vascular tissues.

Observe the horsetail specimens. Note the small narrow leaves at the joints of the stem. Observe the rhizomes and the strobilus (i.e., cone at the stem's apex).

Examine the fern and fern tissues. Look for sori. Sori are collections of sporangia on the undersides of the leaves. Unlike the other seedless vascular plants, ferns have megaphylls.

Observe the slide of the fern prothallium (=prothallus). Know what life stage of alternation of generations it represents.

C. The Gymnosperms

Plants that produce seeds include the gymnosperms and angiosperms. The gymnosperms are characterized by seeds that are not enclosed in ovaries, unlike those of angiosperms. Seeds of gymnosperms are associated with modified leaves called sporophylls which commonly form cones. Both kinds of seed plants are heterosporous in that they produce two types of spores, one to produce the female gametophyte and the other the male gametophyte. Spores develop in the tissues of the sporophyte.

Gymnosperms are classified within the phyla Coniferophyta, Cycadophyta, Ginkgophyta, and Gnetophyta. The Coniferophyta are focused upon here. They include the largest trees in mass (giant sequoia), and some of the tallest (coast redwood) and longest lived (bristlecone pine, giant sequoia, coast redwood).

Phylum Coniferophyta—firs, junipers, larches, and pines

Although distributed throughout the world, conifers dominate the boreal forest (taiga) and alpine regions where temperature, water, and light restrict deciduous tree growth.

Observe the conifer specimen. Compare the life cycle of the pine to the fern.

Examine a cross-section slide of a young pine stem. Locate the tissues presented in Figure 3.1. This stem shows primary growth that provided the initial length and differentiation of tissues in the forms of pith, cortex, and also primary phloem and xylem. Secondary growth that has added girth to the stem includes the periderm, vascular cambium, and

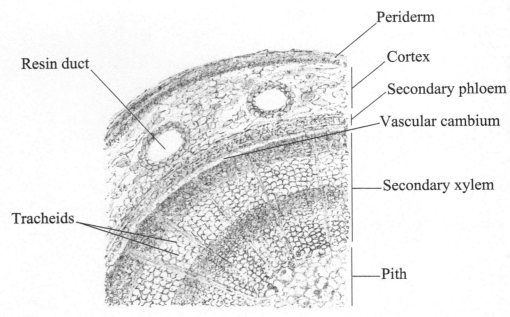

Figure 3.1
Cross section of a young pine *(Pinus)* stem.

most of the phloem and xylem. Secondary phloem is produced externally from the vascular cambium while secondary xylem is produced internally. The primary water and mineral conducting cell of the pine xylem is the tracheid. This cell type also provides support. Pine wood is mostly tracheid in composition. The large resin ducts in the cortex and the smaller resin ducts in the xylem function to deter beetles that attempt to bore into the wood. The sticky resin in the duct contains terpenes and other noxious chemicals.

D. The Angiosperms

Phylum Anthophyta—flowering plants

The major clades of the Anthophyta are:

The Eudicots or dicots—Leaf venation is usually net-like; primary vascular bundles in the stem are in a concentric ring; two cotyledons are present; flower parts are usually in sets of fours or fives; and true secondary growth from vascular cambia occurs.

The monocots—Leaf venation is usually parallel; primary vascular bundles in the stem are scattered; only one cotyledon is present; flower parts are usually in sets of three; and true secondary tissues are absent.

Variation in leaf structure between dicots and monocots is shown by Figure 3.2. Observe the specimens of anthophytes. Identify which are dicots and which are monocots using leaf venation.

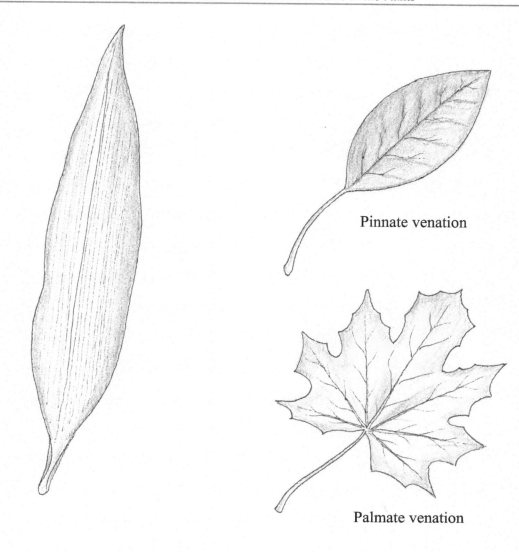

Parallel venation of monocots   Branched venation of dicots

Figure 3.2
Basic difference in leaf venation between dicots and monocots.

The Basic Structure of a Flowering Plant

Figure 3.3 shows the basic anatomy of a flowering plant that is largely lacking woody growth. The plant extends its growth at the terminal buds of the stem and terminal sections of the root. The axillary bud, which is positioned near the petiole of a leaf, can function to replace the leaf or initiate a new lateral stem. Secondary roots emerge from a primary root.

The initial growth of plants, which adds length and differentiation of plant tissues, is called primary growth. Stems, roots, and leaves all show primary growth. In stems and roots, this growth may be followed by secondary growth that is commonly seen through addition of girth and woody tissues.

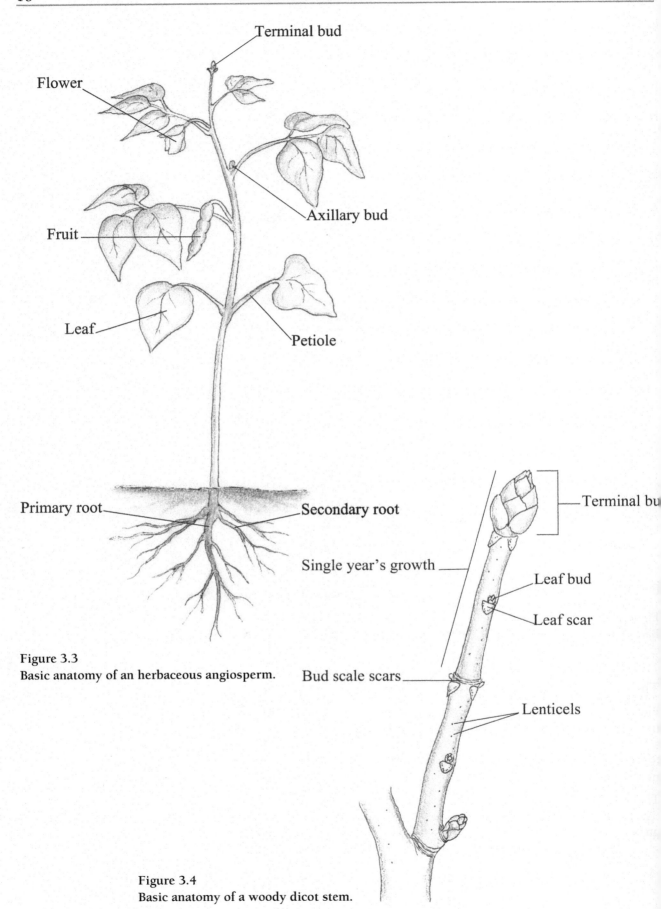

Figure 3.3
Basic anatomy of an herbaceous angiosperm.

Figure 3.4
Basic anatomy of a woody dicot stem.

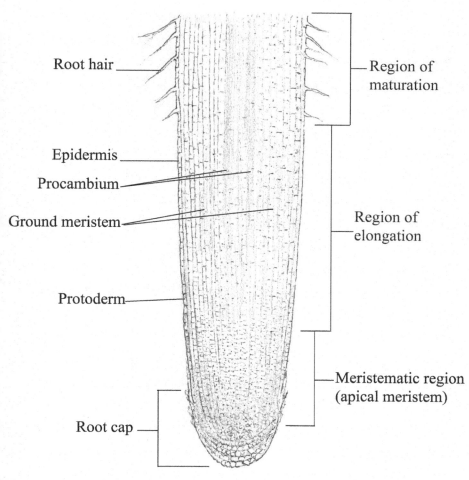

Figure 3.5
Basic anatomy of an angiosperm root tip.

Figure 3.4 illustrates a woody stem that is showing secondary growth. The stem is two years old based on the space between basal and distal bud scale scars (the first year), and between the distal bud scale scars and terminal bud (the second year). Bud scale scars are remnants of buds lost, often at the end of a growing season. Leaf scars are left after leaf detachment.

Observe sample specimens of living plants and also woody stems for the structures identified by Figures 3.3 and 3.4.

The primary growth of a root is illustrated by Figure 3.5. The apical meristem functions to add length to the root. This meristematic region is protected by a root cap. Differentiation of the primary growth tissues occurs behind the apical meristem. The epidermis arises from the protoderm, the primary xylem and phloem from the procambium, and the cortex and pith from the ground meristem. Epidermal cells may show elongated extensions or root hairs which function to increase the surface area for water and mineral absorption.

Continue with a survey of structures of sample flowering plants, examining the three major organs, i.e., the stem, root, and leaf.

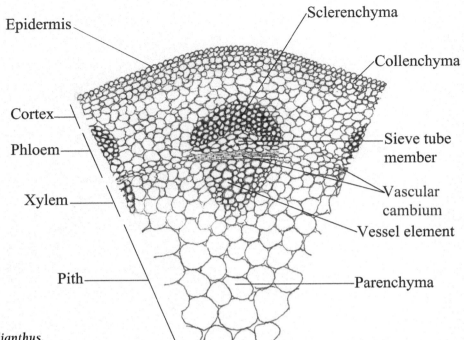

Figure 3.6
Cross section of a stem of *Helianthus*.

The Stem

Examine examples of stems provided. Be able to identify the structures of an herbaceous stem and a woody stem.

Survey the stem cross sections of various anthophytes. Identify and know the functions of the cell and tissues types of each stem. Note primary and secondary growth, and the differences between dicots and monocots.

a. Dicot stems

Helianthus (sunflower) c.s. stem. (Figure 3.6)—This stem shows mostly primary growth. Locate the epidermis, cortex, collencyma, sclerenchyma, phloem, sieve tube member, vascular cambium, xylem, vessel element, and pith. What is the photosynthate conducting cell of the Anthophyta? What relatively large cell in the xylem functions in support and to transport water and minerals?

Tilia (basswood) c.s. stem. (Figure 3.7)—This stem shows secondary growth. Locate the periderm, cortex, sclerenchyma, phloem, sieve tube member, vascular cambium, xylem, vessel element, and pith. A year of growth is indicated by the xylem. In areas where there are seasonal differences in precipitation and/or temperature, the diameters of cells composing the xylem vary. Trees in temperate zones often show their greatest growth during their spring, when favorable environmental conditions occur in precipitation and temperature. Cells of the xylem can grow larger and compose what is known as "early wood". As summer and fall progress, environmental conditions become less conducive to growth, as seen by "late wood" composed of smaller cells. An annual ring of growth consists of a layer of early wood and, to the outside of this wood, a layer of late wood. For the largest cross section on the slide, how old was the stem at the time of cutting?

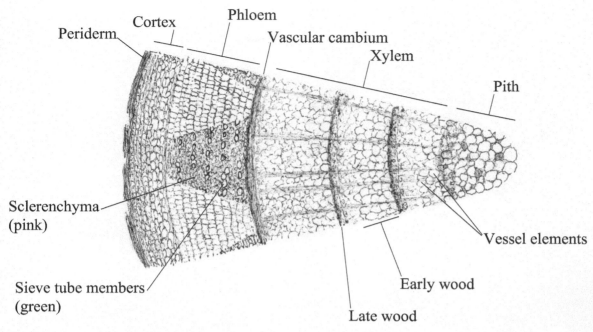

Figure 3.7
Cross section of a stem of *Tilia*.

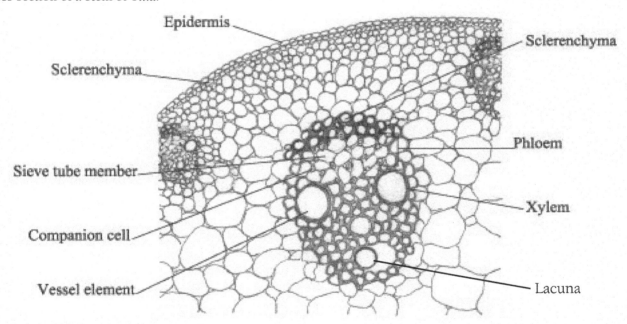

Figure 3.8
Cross section of a stem of *Zea*.

b. Monocot stem

Zea Mays (corn) c.s. stem. (Figure 3.8)—Only primary growth is found in the corn stem. Locate the epidermis, sclerenchyma, phloem, sieve tube member, companion cell, xylem, and vessel element. Look at a vascular bundle as a monkey's face. The two eyes are vessel elements that contribute to the xylem, the forehead contains the phloem with the sieve tube members and tiny companion cells, and sclerenchyma surround the bundle, providing support. The mouth is called the lacuna. It is a space where the

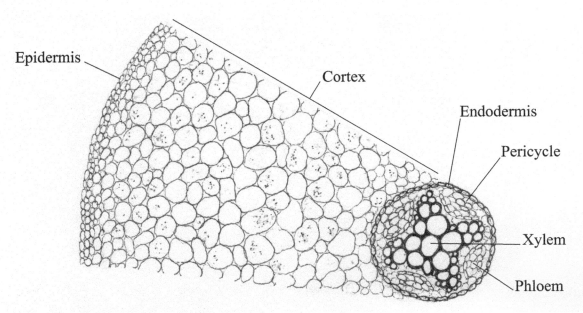

Figure 3.9
Cross section of a root of *Ranunculus*.

first xylem (protoxylem) was once located, but has since been torn apart to create the cavity. Small diameter tracheids located between the vessel elements also compose the xylem.

The Root

Survey the root cross sections of various anthophytes. Identify and know the functions of the cell and tissue types of each stem. Note how the roots differ based on primary and secondary growth, and also between dicots and monocots.

a. Dicot roots

Ranunculus (buttercup) c.s. root. (Figure 3.9)—Note the primary growth. The star-like xylem is a characteristic of an herbaceous dicot. Locate the epidermis, cortex, endodermis, pericycle, phloem, sieve tube member, xylem, and vessel element.

Tilia (basswood) c.s. root. (Figure 3.10)—Note the secondary growth. Locate the periderm, sclerenchyma, phloem, sieve tube member, vascular cambium, xylem, and vessel element. What has happened to the cortex?

b. The monocot root

Zea Mays (corn) c.s. root.—large round xylem that forms a ring in cross section of root.

The Leaf

Survey the variation shown in leaf structure among samples available.

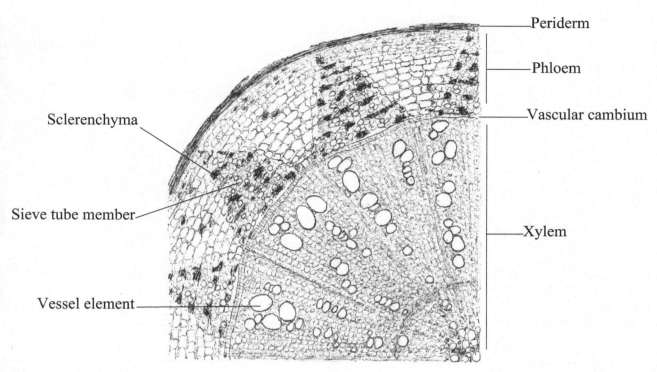

Figure 3.10
Cross section of a root of *Tilia*.

a. The dicot leaf

Syringa (lilac) c.s. leaf. (Figure 3.11a)—Locate the epidermis, palisade mesophyll, spongy mesophyll, xylem, phloem, and stomata (stoma is singular). The presence of a differentiated mesophyll is also characteristic of a C3 plant.

b. The monocot leaf

Zea (corn) c.s. leaf. (Figure 3.11b)—Find a major or thicker vein. The large cells in the vein are vessel elements that compose xylem. Locate the greenish phloem immediately below the xylem. In the phloem, see if you can differentiate the sieve tube cells from the companion cells in the phloem. Identify the xylem and phloem. Identify the well-defined bundle sheath around the vascular bundle which is an identifying feature of a C4 plant.

c. Leaf variation according to habitat

Mesophytic, hydrophytic, and xerophytic plant leaves c.s.—this slide contains leaves of plants that live in 3 contrasting environments. A mesophytic plant lives in terrestrial environments that receive moderate amounts of precipitation, a hydrophytic plant lives in water, and a xerophytic plant lives in dry environments. Identify the mesophytic leaf (Figure 3.11a) noting that the palisade mesophyll (palisade parenchyma) is about equal in depth as the spongy mesophyll (spongy parenchyma) along a blade when viewing a

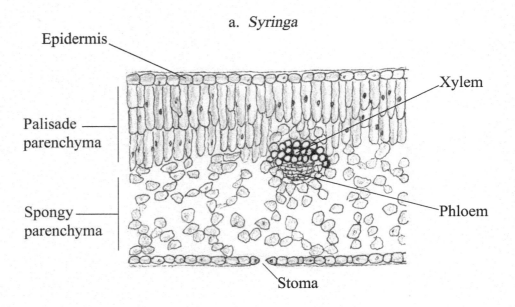
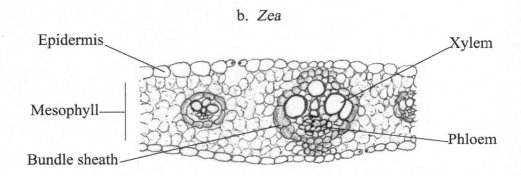
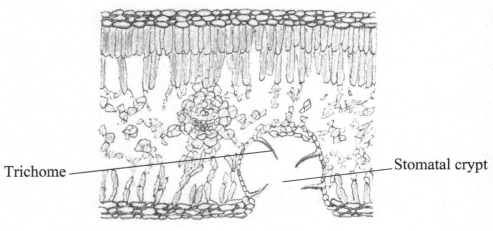

Figure 3.11
Leaf cross section of a. *Syringa*, b. *Zea*, and c. a xerophytic plant.

distance away from the central vein. The spongy mesophyll of a hydrophytic leaf is deeper with large air spaces formed by aerenchyma. Aerenchyma create buoyant air spaces. The xerophytic leaf (Figure 3.11c) is characterized by stomatal crypts along the blade. Stomata are found within the crypts. Trichomes are hair-like extensions of cells that block the loss of water from stomata in the stomatal crypts.

Flowers

Flower structure is very diverse among species of Anthophyta as affected by the environment, particularly the means of pollination. Wind-pollinated flowers are commonly without color and scent. These flowers are often clustered together in larger numbers and may be suspended in tassels as to shed pollen. In contrast, flowers requiring animal pollination are often colorful and/or with scent to attract the pollinators. Most animal pollinators are insects (primarily ants, bees, beetles, flies, moths, and butterflies), but can also include vertebrates like bats, birds, and rodents. The plant and animal pollinator often have a highly co-evolved relationship.

Flowers may occur singly or in a cluster called an inflorescence and are held on a stalk called the peduncle. In an inflorescence, the peduncle branches into pedicels that hold individual flowers. The base of the flower, the receptacle, begins as a swelling on the peduncle or pedicel. Floral parts are derived form highly modified leaves that occur as a series of whorls or encircled layers attached to the receptacle. The outermost whorl is the calyx comprised of sepals, followed by the corolla comprised of petals, and then reproductive parts. Male gametes are carried in pollen grains produced in a stamen, a modified microsporophyll. Most stamens consist of an elongated stalk, the filament, which supports an anther. Each anther has four sporangia which produce microspores from which the pollen grain (or male gametophyte) develops. Female gametes are produced in female gametophytes located in an ovule or ovules of a carpel. Four megaspores are produced in each ovule by meiosis, but only one survives. This megaspore gives rise to a female gametophyte.

The number of sepals, and also petals, is typically 3 for monocots, and 4 or 5 for eudicots. Perfect flowers have pistils derived from one or more carpels, and stamens (Figure 3.12). The terms carpel and pistil may be used synonymously when the pistil is comprised of a single carpel. A pistil includes a stigma that receives the pollen, a style down which the germinating pollen grows, and an ovary which can contain one or many ovules. Each ovule includes a female gametophyte. The stamen consists of a filament that suspends the anther. The anther bears the pollen. Pollen of angiosperms contains a tube nucleus to support pollen growth to the ovary and two sperm cells, one to fertilize the egg cell and the other to fertilize a central cell with polar nuclei, producing nutritive endosperm. Fertilized ovules develop into seeds.

Flowers may be imperfect (unisexual) when they are either carpellate or staminate. Flowers of monoecious plants can be either imperfect or perfect (having both carpels and stamens). If a monoecious plant possesses imperfect flowers, both carpellate and staminate flowers must be present on the plant. Dioecious plants, by definition, can only have carpellate flowers or staminate flowers.

View a cross section of a *Lilium* ovary. Locate and count the ovules in the center of the flower.

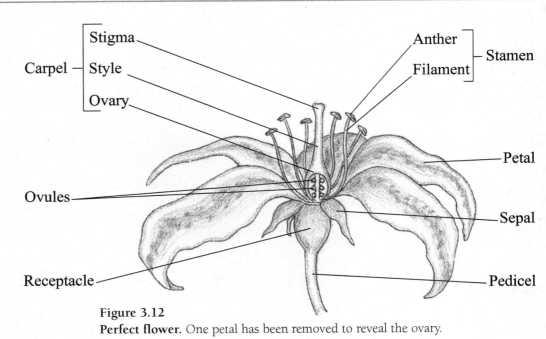

Figure 3.12
Perfect flower. One petal has been removed to reveal the ovary.

Observe various models of flowers. Locate the individual parts of a flower described earlier.

If available, examine fresh flowers. Dissect the separate parts of the flower, beginning with the outermost whorl and working inward. Arrange the floral parts in their respective whorls on paper. Carefully splice open ovaries to locate the ovules. Using tape, mount and label your arranged flower parts on paper.

How might you tell if the ovary is derived from the fusion of more than one carpel? Can you predict how each type of flower is pollinated? If animal pollinated, can you predict the identity of the pollinator?

Fruits

A fruit develops from the ovary of a flower and often accessory structures. Fruits are a means by which flowering plants disperse their seeds. Fruits can be dispersed explosively (e.g., impatiens), by wind (e.g., dandelions), water (e.g., coconuts), or animals (e.g., cherries). Seeds can stick to and be dispersed on the fur and feathers of animals, or be ingested and dispersed when the animal defecates.

Fruits are classified in a variety of ways, such as how the fruit develops, if the fruit is dry or fleshy, and if the fruit dehisces (splits open) at maturity. A simple fruit develops from one pistil (carpel) of a single flower (e.g., bean). An aggregate fruit develops from separate pistils of a single flower (e.g., raspberry or blackberry). A multiple fruit develops from pistils of many different flowers in an inflorescence (e.g., pineapple). Some fruits contain not only the ripened ovary, but also tissues that develop from other parts of the flower (e.g., an apple's fleshy portion develops from the receptacle of the flower).

Survey the variation in fruit structure shown by the Anthophyta. Understand fruit structure in relation to the life cycle of an angiosperm and means of seed dispersal.

Cut open a variety of fruits. Locate the ovary wall, seeds, and any other remains or derivatives of flowers from which they develop.

EXERCISE
Classification and Structure of the Plants

Name_____ Date_____

Domain Eucarya—Kingdom Plantae

1. The mosses tend to be small in size. From the viewpoint of structure, what limits how large a moss can be? What advantage is smallness to a plant like a moss?

2. How does the protonema fit into the moss life cycle?

3. What are the functions of the antheridia and archegonia?

4. For *Anthoceros,* what is the function of the tissue just inside the epidermis?

5. What tissues characterize a vascular plant?

6. What is the dominant life stage of the vascular plants?

7. What is the dominant organ of the fern?

8. What is the function of the strobilus of the horsetail?

9. What life stage of the fern does the prothallium represent?

10. What is the major difference between microphylls and megaphylls?

11. Where does the pinecone fit into the life cycle?

12. Contrast dicot and monocot stem and root structures based on primary growth, secondary growth, and the arrangement of the vascular tissues.

13. What are early wood and late wood? How can the age of a tree be determined?

14. Leaves of plants vary greatly in design, shape, and even color. What selective forces act upon plants to affect leaf design?

15. If a leaf is vertically oriented, would you expect its mesophyll to be divided into spongy and palisade layers? Explain.

16. Describe adaptations that can prevent a perfect flower from self-pollinating. Explain potential advantages and disadvantages of self pollination.

17. Describe a co-evolutionary relationship between a particular flowering plant and its pollinator. Be sure to explain how traits in one are influenced by the other.

18. Distinguish between pollination and fertilization.

19. Explain how seedless varieties of fruits, such as seedless grapes and watermelons, are propagated.

20. Explain how seeds of an apple may be dispersed in nature.

Discussion

Coast redwoods (*Sequoia sempervirens*) grow in temperate rainforests along northern California and Oregon. They can exceed 110m in height, placing them among tallest of the trees. The highest tips of tall trees often show evidence of wilting. What limits how tall trees can grow and why are the redwoods capable of being so tall?

4 Classification of the Animals

Domain Eucarya—Kingdom Animalia

Animals are believed to have evolved 900 mya (million years ago) based on trends in DNA change over time. As marine organisms, the first animals took advantage of a larger size in capturing prey and avoiding predation. However, as with plants, a larger body size necessitated supportive tissue which later organized into organs and organ systems. By the Cambrian explosion 543 to 490 mya, most of the major animal phyla had evolved. This period saw a large diversification of aquatic invertebrates and also those terrestrial. By 520 mya, the vertebrates were evolving and by 400 mya, the first tetrapods. Hominids, which include us, are recent arrivals, originating some 7 mya.

The Animalia includes some 10 million to 200 million species, of which a little over 1 million species have been identified. Almost all animals (over 99%) are invertebrates. Most are small and/or marine, explaining the lack of taxonomic treatment given to them. Animals are chemoheterotrophic eukaryotes, multicellular, and most show some degree of tissue and organ development. The phyla are organized beginning with those that are structurally less complex and proceeding to those more complex. Remember, evolution does not necessarily follow this path, so simplification may be adaptive to various life styles (e.g., internal parasites). As you survey the following animals, read your lecture text concerning what tissues and organs each phylum possesses and the selective advantages to these "complexities" in view of how the animals exist in nature.

Know the taxonomy of each of the following groups of animals as given and be able to identify the various structures. Answer the questions on the data sheet.

Phylum Porifera—the sponges

The sessile sponges lack organ and tissue differentiation. Mostly marine, but with some freshwater species, sponges are suspension feeders that draw organic material as food from water. Sponges have an outer and inner layer of cells imbedded in a gelatinous mesoglea. The most basic structure of a sponge resembles a paper bag with holes (ostia) on the side. Water is drawn through the ostia by flagellated choanocytes (choano = collar, cyte = cell) which line an inner cavity known as the spongocoel. The water then exits through the osculum. Organic matter is captured by cells lining the spongocoel and distributed by mobile cells called amoebocytes to other cells in the body. Many sponges have siliceous or calcium carbonate spicules that mesh together to provide skeletal support and also function to deter predators. Instead of spicules, other sponges, like bath sponges, have protein fibers called spongin.

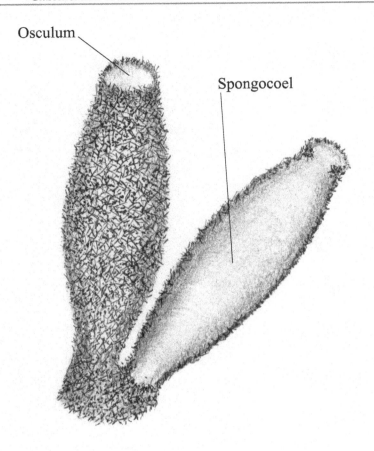

Figure 4.1
Longtitudinal section of a simple sponge.

Examine the representative sponge specimens.

Slide of longitudinal cross section of *Grantia* or *Scypha* (c/s)—identify the spongocoel and osculum (Figure 4.1).

Examine the slide of spicules.

Phylum Cnidaria—hydra, Portuguese man-of-war, jellyfishes, corals, sea anemones, and sea fans

Cnidarians are diploblastic, meaning they form from two embryonic germ layers: endoderm and ectoderm. However, they show extracellular digestion inside a central cavity, called the gastrovascular cavity, that has one opening to serve as both mouth and anus. While some are sessile, others move by gliding, somersaulting, floating, and even rudimentary swimming. Cnidarians have stinging cells called cnidocytes. These cells contain nematocysts which can project a barbed thread. Upon penetration of the prey, the thread ejects a small amount of toxin. Other features include a nervous system called a nerve net, so named because it resembles a fishing net. Some cnidarians even have simple photoreceptors that sense light. Many cnidarians show alternation of generations which includes a sessile and asexual polyp and a motile sexually reproducing medusa. One of these life stages tends to dominate depending on the type of cnidarian. Coral only exist as polyps.

Examine the representative specimens of cnidarians.

Class Hydrozoa—hydroids

Observe the *Hydra* slide and, if available, living material. Identify the tentacles, gastrovascular cavity and mouth. If available, note the behavior shown by the live *Hydra*.

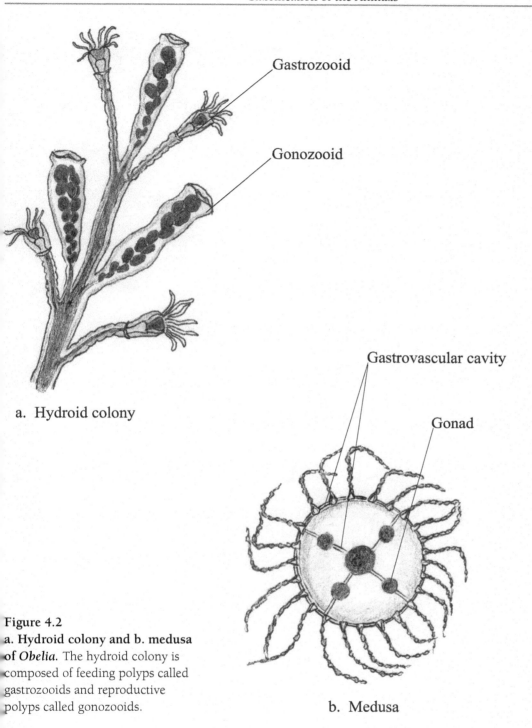

Figure 4.2
a. Hydroid colony and b. medusa of *Obelia*. The hydroid colony is composed of feeding polyps called gastrozooids and reproductive polyps called gonozooids.

Obelia slides. This is a colonial hydrozoan. Viewing the slide of a polyp colony, differentiate the gastrozooids/feeding polyps (with tentacles) from the gonozooids/reproductive polyps (without tentacles) (Figure 4.2). In the slide of a medusa, locate the gastrovascular cavity and four gonads.

Examine a specimen of Portuguese man-of-war. What appears to be a single specimen is actually a hydroid colony.

Class Scyphozoa—the jellyfishes

Jellyfishes are marine predators which show a weak ability to swim. Some can be quite large with tentacles extending several meters.

Class Anthozoa—corals, sea fans, and sea anemones

The anthozoans do not show a medusoid life stage. All are marine. Coral have particular ecological importance in creating coral reefs. The Great Barrier Reef off Northeastern Australia extends some 6000 km. Reefs are highly biodiverse, serving as habitats where many other organisms come to reproduce.

Phylum Ctenophora—comb jellies

Comb jellies are marine animals whose name comes from the eight rows of cilia that look like combs. The comb jellies are diploblastic and have a gastrovascular cavity like the Cnidaria. However, they lack cnidocytes, and instead capture tiny prey with two sticky tentacles. Comb jellies also have a complete gut with two anal pores that lead from the gastrovascular cavity. Some are invasive, having been introduced from the Americas to the Black and Caspian seas from ship ballasts.

Observe the representative specimen of comb jellies.

Phylum Platyhelminthes—the flatworms

The flatworms are believed to be the first motile predators. They are triploblastic, as are the remaining animals covered in this laboratory exercise. They have tissues and organs derived from three germ layers: ectoderm, mesoderm and endoderm. These acoelomate worms show bilateral symmetry, which is more adaptive to organized movement in comparison to the radial symmetry of the cnidarians. They have an incomplete digestive system and an excretory/osmoregulatory system consisting of protonephridia. Each protonephridium has two lateral canals which are terminated by ciliated flame cells. These cells collect water and wastes, directing them up and out of the lateral canals. Free-living flatworms like planarians show the beginnings of cephalization where nervous tissue is concentrated in the anterior end as two cerebral ganglia. These ganglia, in part, support photoreceptors.

Examine the representative specimens of flatworms.

Class Turbellaria—planarians

This class contains the free-living flatworms. Most are marine. Planarians are predators and scavengers which can be commonly found in local freshwater streams and lakes.

Observe prepared whole mount (w/m) slides of planaria, and if available, live specimens. Locate the ocelli, intestine, and pharynx (Figure 4.3). Ocelli contain photoreceptors.

Class Trematoda—flukes

Flukes are internal parasites of vertebrates. They show complex life cycles that may include several intermediate hosts harboring the non-sexual life stages. Vertebrates provide the definitive host in which the flukes become sexual.

Observe w/m slides of *Fasciola* and the cercaria. *Fasciola* represents the sexual life stage of a fluke, while the cercaria are larval forms of the sexual stage which infest the definitive host.

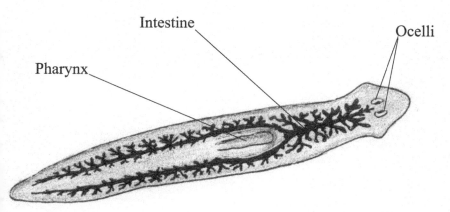

Figure 4.3
Whole-mount view of a planarian.

Class Monogenea—fish flukes

Fish flukes are largely ectoparasites of marine and freshwater fish, but they also parasitize amphibians, reptiles, and cephalopods (squids and octopuses). Development proceeds from the egg, to two larval stages, and finally the adult that will seek a host to parasitize. The adult attaches to the host with a haptor, a circular attachment disk with hooks.

Observe a w/m slide of *Gyrodactylus* and locate the haptor.

Class Cestoda—tapeworms

Tapeworms are internal parasites of vertebrates. They have one intermediate host, which can be a vertebrate, but also includes insects. Tapeworms lack a digestive system, and based on the tissues that they can inhabit, show a complex life cycle. The anterior segment of the tapeworm is called the scolex. This attachment organ produces proglottids which are massive reproductive organs. As the proglottids mature, they, and the eggs they produce, are shed from the body and through the feces of the host. Eggs or proglottids then are consumed by an intermediate host.

Examine a w/m slide of *Taenia* and locate the scolex and proglottids (Figure 4.4).

Examine the hydatid cysts infusing the rat liver. Each cyst harbors a larval tapeworm called a cysticercus.

Phylum Rotifera—rotifers or wheel animals

Rotifers are tiny pseudocoelomate animals that range from 0.1 to 1 mm in length. The alternative name of wheel animals comes from the anterior corona. The corona consists of two circular structures surrounded by cilia. When the cilia move, the resemblance is that of two spinning wheels. The corona is used in movement and to catch food. Rotifers are pseudocoelomate, have a complete gut, a pair of protonephridia for osmoregulation and secretion, and even photoreceptors. The animals are known to be eutelic, having constancy in cell number per tissues and organs. Most live in freshwater, with some being marine. They tend to be predators, although a few are parasitic.

Observe a w/m slide containing rotifers.

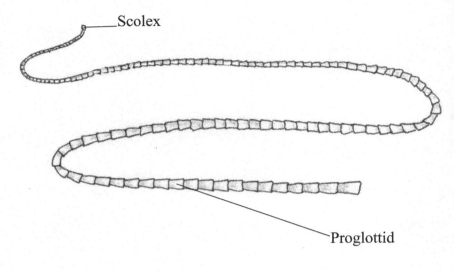

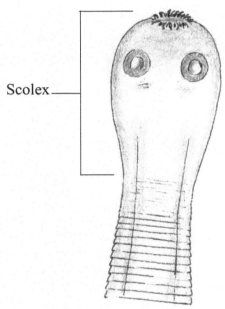

Figure 4.4
Whole-mount view of a tapeworm.
Proglottids develop behind the scolex.

Phylum Ectoprocta—bryozoans or moss animals

The eucoelomate bryozoans superficially resemble cnidarian polyps, but have a U-shaped digestive system with mouth and anus. Bryozoans are characterized by a lophophore, comprised of ciliated tentacles that surround the mouth. These structures function to guide food to the mouth. Many bryozoans are colonial. They inhabit streams and lakes, but most are marine, where many contribute to reef building.

Observe a w/m slide of bryozoans.

Phylum Brachiopoda—brachiopods or lamp shells

Two shells enclose the body of a brachiopod, giving it the appearance of a bivalve. However, the shell halves are oriented dorsally and ventrally, not laterally as in clams. Brachiopods have a lophophore, coelom, and a U-shaped digestive system like the Ectoprocta. They show a more abundant past, with 30,000 extinct species currently described, compared to the 335 species surviving today. All are marine.

Phylum Mollusca—shelled animals

Mollusks are well represented in the fossil record as their shells help in preservation and fossilization. With some 110,000 extant species, another 35,000 extinct species have been identified from the fossil record. A thin and soft membrane, the mantle, secretes the shell. Not all mollusks have shells, as a shell can restrict movement through tight areas. Many members of this eucoelomate phylum have gills to facilitate gas exchange. Tube-like metanephridia serve as excretory organs. Most have an open circulatory system where hemolymph ("blood") intermixes with interstitial fluids. A heart in many of these mollusks forces hemolymph through the sinuses, which are found between organs. Cephalopods, which include squids and octopods, have a closed circulatory system where blood is separated from interstitial fluids by being confined in blood vessels. The mollusks exhibit a range of trophic roles to include predators, herbivores, scavengers, and even a few parasites.

Examine the representative specimens of mollusks.

Class Gastropoda—snails, limpets, and slugs

The gastropods show the greatest degree of adaptive radiation among mollusks, with a few species inhabiting terrestrial environments. Adaptations of terrestrial species include excretion of uric acid in contrast to ammonia, which is the major nitrogenous product of freshwater and marine species.

Class Polyplacophora—chitons

Chitons are marine mollusks that typically bear eight overlapping transverse plates that serve as their shell. Some have reduced plates or no plates at all. Chitons inhabit rocky shoreline areas, where they graze algae growing on rocks. A muscular foot helps chitons adhere to rock in tidal zones.

Class Bivalvia—mussels, scallops, oysters, and clams

Bivalves are so named based on the two shells that compose their bodies. These laterally compressed animals commonly filter organic material from water as their source of food. They either partially bury themselves into substrate or attach themselves to rocks and other immovable objects to feed. Some have a parasitic life stage which infests fish. Members of the genus *Entovalva* inhabit the guts of sea cucumbers. Other bivalves bore into sandstone, coralline rock, wood, and other submerged objects. Bivalves inhabit both marine and freshwater environments. They are often fairly common in relatively unpolluted streams with little sediment load. Shells reflect seasonal changes in the environment, creating annual rings. Based on these rings, some clams in the North Atlantic are believed to be 150 years old.

Dissect a clam. Use Figure 4.5 to help locate the various parts. First determine the orientation, i.e., the anterior, posterior, and lateral sides to the clam. Locate the exterior umbo (on dorsum) and growth lines on the shell.

To open the clam, take a scalpel and carefully insert it into the space between the shells. Cut through the anterior and posterior adductor muscles which function to control the shells. Identify the thin mantle that lines the inner sides of the shells. Should an object like a grain of sand be captured in the folds of the mantle, the object may be surrounded by calcium carbonate to form a pearl. Locate the muscular foot that aids in locomotion and burrowing.

Locate the incurrent and excurrent siphons at the posterior end. They draw water in and out of the clam, respectively. Cilia lining the soft tissue inside the shells transport food to the mouth at the anterior end. Locate the palp, mouth, anterior retractor muscle which controls the foot, and gills. The tube-like intestine is found just inside the umbo. Look for the heart, which forms a bulge on the intestine. You will next need to slice through the foot and visceral mass (which contains the organs). Locate the stomach, liver, intestinal tract, and gonads.

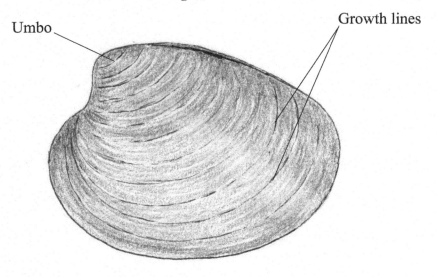

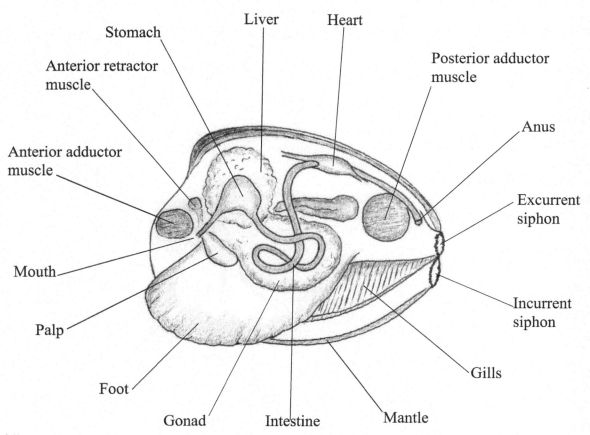

Figure 4.5
External and internal views of the clam.

Class Cephalopoda—nautili, squids, and octopods

The cephalopods include some of the most intelligent invertebrates. They show evidence of problem solving and observational learning. Strictly marine, the cephalopods are camouflage artists, capable of changing their body colors to match the background in order to escape predators and capture prey.

Squid dissection

Use Figure 4.6 to help locate the body structures. Before doing any cutting, locate the anterior and posterior arms, mouth, eyes, siphon, mantle, and tentacles. In locating the arms and mouth, understand the orientation of the squid (i.e., the anterior, posterior, ventral, dorsal, and lateral sides). Slice through the mantle and head, cutting on the posterior side. Locate the pen (remnant of the shell), mouth, buccal bulb, esophagus, brain that surrounds the anterior esophagus, anus, rectum, stomach, cecum, gills, branchial hearts that emerge from the base of both gills, systemic heart that lies somewhat between the gills, and gonad (if present). Food is passed into the mouth, through the buccal bulb and esophagus to the stomach and cecum where digestion occurs. The branchial hearts send blood through the gills to be oxygenated. The systemic heart pumps the oxygenated blood to the body.

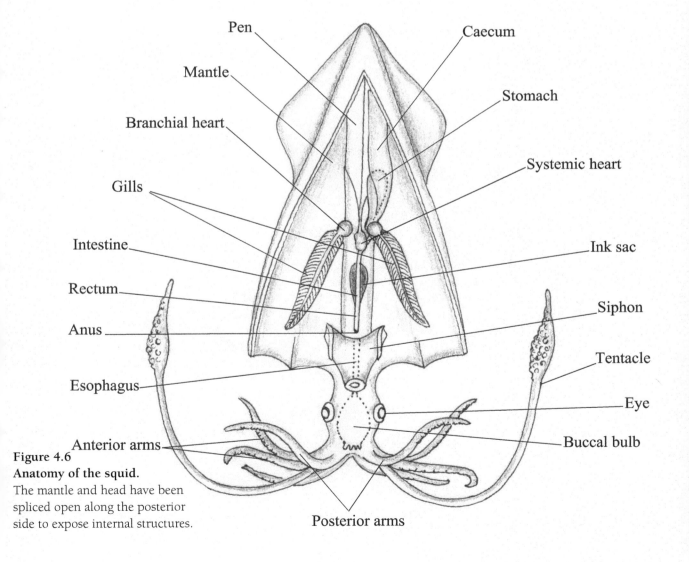

Figure 4.6
Anatomy of the squid.
The mantle and head have been spliced open along the posterior side to expose internal structures.

Phylum Annelida—segmented worms

Genetically, this eucoelomate phylum is most closely related to the Mollusca. They also share a similar form of embryonic development, a ciliated larval stage called a trochophore larva, and metanephridia for excretory organs. Unlike most mollusks, the worms have a closed circulatory system with five pairs of pumping vessels surrounding the esophagus. Segmentation is clearly apparent in the epidermis, underlying muscle, and the pair of metanephridia per segment. Within the coelom, segmentation stops. The fluid-filled coelom provides a hydrostatic skeleton.

Examine the representative specimens of segmented worms.

Class Polychaeta—the sandworms

Most sandworms are characterized by feet-like lateral appendages called parapodia with hair-like setae. Parapodia help in motility and gas exchange. While many polychaetes burrow in sand, functioning as suspension feeders, others are active predators and scavengers. All are marine.

Class Clitellata—the earthworms and leeches

Members of this class have an external structure called a clitellum (Figure 4.7) that secretes mucus to create a protective cocoon around the eggs.

Subclass Oligochaeta—the earthworms

Earthworms are terrestrial and aquatic. They are known to be important soil mixers, aerators, and decomposers. While valued by farmers, earthworms introduced to North America by European colonizers have shown to be invasive and damaging to native communities by altering the soil environment.

Dissect an earthworm. Use Figure 4.7 to help in identifying the structure. First locate the external mouth, male genital pore and clitellum. Carefully cut along the dorsal side of the worm as identified by the darker side, which is due to a dorsal blood vessel. Only cut through the epidermis and underlying muscle, leaving the digestive system untouched. Use dissection pins to spread the worm open. Do not place a pin at the anterior tip as this will likely damage the supra-pharyngeal ganglia. These ganglia appear as two pin-sized white dots that are attached to one another. Locate the supra-pharyngeal ganglia, pharynx, esophagus, crop, gizzard, intestine, circum-pharyngeal commissures, large whitish seminal vesicles, white-colored ventral nerve cord, metanephridia, pseudohearts (or pumping vessels that surround the esophagus), dorsal blood vessel, and septa. The ventral nerve cord splits ventrally under the pharynx into two circum-pharyngeal commissures which fuse on the dorsal tip of the pharynx to form the two suprapharyngeal ganglia (brain). The muscular pharynx takes in food which travels through the esophagus to a storage structure, the crop. The muscular gizzard grinds the ingested food, while digestion and assimilation occur in the intestine.

Subclass Hirudinea—leeches

Leeches are parasitic to predatory. They lack bristles or parapodia, but have a posterior sucker. They are commonly found in freshwater habitats but are marine or terrestrial.

Phylum Nematoda—nematodes or round worms

Nematodes are closely related to the Phylum Arthropoda. Together, these phyla are among the most successful animals in terms of diversity and

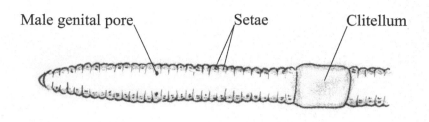

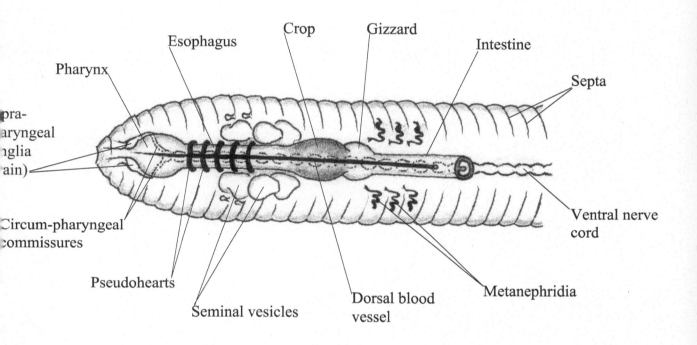

Figure 4.7
External and internal view of the earthworm.

abundance. Like that of the arthropods, the body of a nematode is covered by a protective acellular cuticle. The worm must molt (a process known as ecdysis) in order to increase in size. Nematodes have a pseudocoel which serves as a hydrostatic skeleton and circulatory system. Two simple tubules function in excretion. Most nematodes are free-living, exploiting aquatic and terrestrial environments as herbivores, predators, and decomposers. Some plague humans as parasites, to include *Ascaris* which you are to dissect. An estimated one billion people are infested by *Ascaris* which inhabits the intestine of its host in the adult stage. Eutelic, nematodes show constancy in cell number in most of the tissues of the adult, making them ideal study subjects for understanding animal development.

Examine the representative specimens of round worms.

Dissect male and female specimens of *Ascaris lumbricoides* using Figure 4.8 as a guide. Make a longitudinal cut with a scalpel down the length of the worm. Be careful to only cut through the outer layer. Use dissection pins to spread the

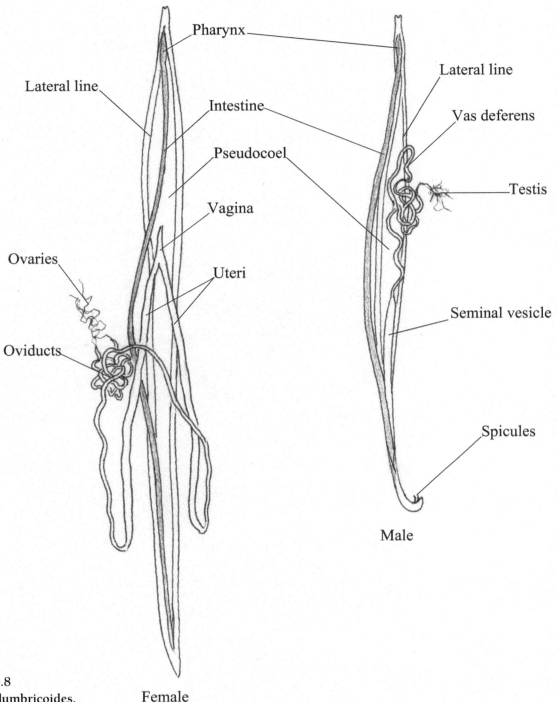

Figure 4.8
Ascaris lumbricoides.

worm open. Locate the pharynx, intestine, lateral line, pseudocoel, ovaries, oviducts, uteri, vagina, testes, vas deferens, seminal vesicles and spicules.

Phylum Onychophora—onychophorans or velvet worms

Velvet worms are named based on the hair-like structures that cover their bodies. They have survived since prehistoric times first as marine species and later as terrestrial species in warm, humid environments. Eucoelomates, they are closely related to the Arthropoda and largely predatory.

Examine the representative specimens of velvet worms.

Phylum Arthropoda—trilobites, chelicerates, myriapods, crustaceans, and insects

Arthropods are the most abundant above-ground animals on earth, and are represented by over one million species, most of which are insects. They are highly segmented, where segments are fused as functional units called tagmata. Insects, for example, have three major tagmata as adults, those being the head, thorax, and abdomen. All arthropods are eucoelomates and have an open circulatory system with a tubular heart typically located in the abdomen. Gas exchange organs reflect the environment the arthropod inhabits. Aquatic arthropods, to include most crustaceans, have gills. Many chelicerates have an invagination of the ventral abdominal wall called a book lung. Insects have a tubular tracheal system with openings to the outside largely arranged along the lateral side of the abdomen. Excretion occurs through metanephridia among some arthropods and filamentous malpighian tubules in insects. The digestive systems of arthropods are diverse and often complex as to enable digestion of difficult materials, typically with the aid of endosymbiotic bacteria and protozoans. In many arthropods, antennae and even the hair-like setae found on the body are geared for sensory function.

Examine the representative specimens of arthropods.

Subphylum Trilobita—trilobites

Trilobites were very common in marine habitats between 500 and 250 million years ago when they suddenly disappeared. They include benthic and pelagic species that possessed antennae, compound eyes, and legs.

Subphylum Chelicerata—spiders, mites, ticks, scorpions, pseudoscorpions, and horseshoe crabs

Chelicerates are characterized by claw-like or fang-like head appendages called chelicerae. These head appendages are followed by pedipalps that can also resemble claws or are leg-like.

Class Merostomata—horseshoe crabs

These living fossils represent a greater diversity of species of the prehistoric past. Horseshoe crabs inhabit coastline areas of eastern North America, Southeast Asia, and Indonesia. The hemolymph of horseshoe crabs has pharmaceutical use in testing for contamination by bacterial endotoxins and screening for bacterial diseases such as meningitis and gonorrhea. Their eggs provide a critical resource for resident and migrating shorebirds. The survival of horseshoe crabs is under threat by anthropogenic activities to include oil spills, pollution, and harvesting them as eel bait.

Examine the horseshoe crab specimen. Locate the small pair of chelicerae nearest the mouth, the adjacent pair of pedipalps, and the four pairs of walking legs.

Class Arachnida—spiders, mites, ticks, scorpions and pseudoscorpions

The first two pairs of head appendages on the arachnid are the chelicerae and pedipalps. These are followed by four pairs of walking legs. The pedipalps may be leg-like and confused as such. Antennae are absent. The body consists of two tagmata, a cephalothorax (fused head and thorax) and abdomen.

Subphylum Myriapoda

All myriapods have one pair of antennae, mandibles, and many pairs of legs to help propel through the substrate in which they live.

Class Diplopoda—millipedes

Millipedes are mostly detritivores, consuming dead plant matter. They have two pairs of legs per segment. Leg pairs may number approximately one hundred.

Class Chilopoda—centipedes

Centipedes prey largely on other arthropods. They are relatively quick moving, having one pair of legs per segment, 15 or more pair of legs, and a flattened body.

Subphylum Crustacea—crayfishes, lobsters, crabs, copepods, shrimp, barnacles, and cladocerans

Crustaceans have two pairs of antennae, mandibles, and commonly, two major tagmata, the cephalothorax and abdomen. They include predators, scavengers, and suspension feeders. While most are aquatic, a few are terrestrial.

Dissect a crayfish. Use Figures 4.9 and 4.10 to help identify the structures. First, identify the following external structures: cephalothorax, carapace, abdomen, antennae, compound eyes, and mandibles.

Expose the internal parts of the crayfish by removing the dorsal sections of the carapace and abdominal segments. Slip a pair of scissors into the posterior and lateral side of the carapace. Cut the carapace up to and behind the compound eyes. Continue to cut down the opposite side of the carapace. Carefully connect the lateral cuts at the posterior end of the carapace by cutting across the dorsum. Lift the carapace, making sure to detach all soft connecting structures from the carapace as you do so. Next cut the top cuticle of the tail. Remove the dorsal muscles that cover the intestine.

Identify the stomach, digestive gland, gills, heart (which is located dorsally between the gills), ostia (or holes in the heart that draw in hemolymph), and intestine. The stomach functions to mix and digest food, and the digestive gland provides enzymes, and stores glycogen, fat, and calcium. Assimilation occurs beginning with the stomach and on into the intestine. Lift up the stomach and remove it. Be able to identify the brain and antennal glands. The antennal glands, which are modified metanephridia, lie inside the head behind each compound eye.

Subphylum Hexapoda—springtails and insects

The hexopods are characterized by adults having three pairs of legs and usually one pair of antennae.

Class Insecta—the insects

The largest class of all life forms is the class Insecta. Insects range across the nutritional spectrum to include predators, parasites, detritivores, and herbivores. Many have major economic and medical significance as major consumers of plants and animals, and vectors of disease. Many more have

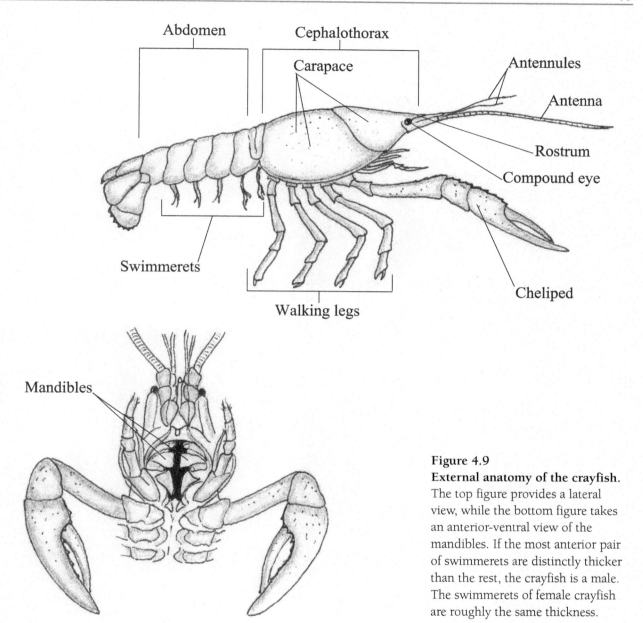

Figure 4.9
External anatomy of the crayfish. The top figure provides a lateral view, while the bottom figure takes an anterior-ventral view of the mandibles. If the most anterior pair of swimmerets are distinctly thicker than the rest, the crayfish is a male. The swimmerets of female crayfish are roughly the same thickness.

invaluable importance as pollinators, a source of food, control agents of pests that include other arthropods, and regulators of community productivity. Insects include the only arthropods that have wings.

Be able to identify hemimetabolous development and holometabolous development. Hemimetabolous development is characterized by three major life stages: the egg, nymph (if aquatic, then naiad), and adult. The nymph is the primary growth stage where nymphs molt to become larger nymphs until the adult stage is reached. Nymphs resemble adult stages minus reproductive structures and wings, should the adult stage have wings. Holometabolous development is characterized by four major life stages: the egg, larva, pupa, and adult. Again the larva is the major growth stage, becoming larger through molts until pupating. Larvae look unlike the adults and often feed on different foods as to avoid competition with adults. Most insects, by number of species, show holometabolous development.

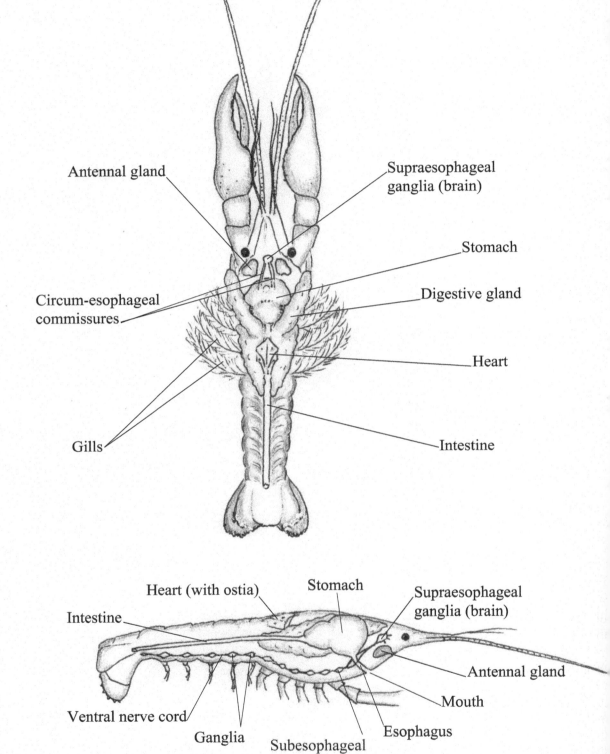

Figure 4.10
Internal anatomy of the crayfish. The top figure provides a dorsal view and the bottom figure provides a lateral view with organs exposed.

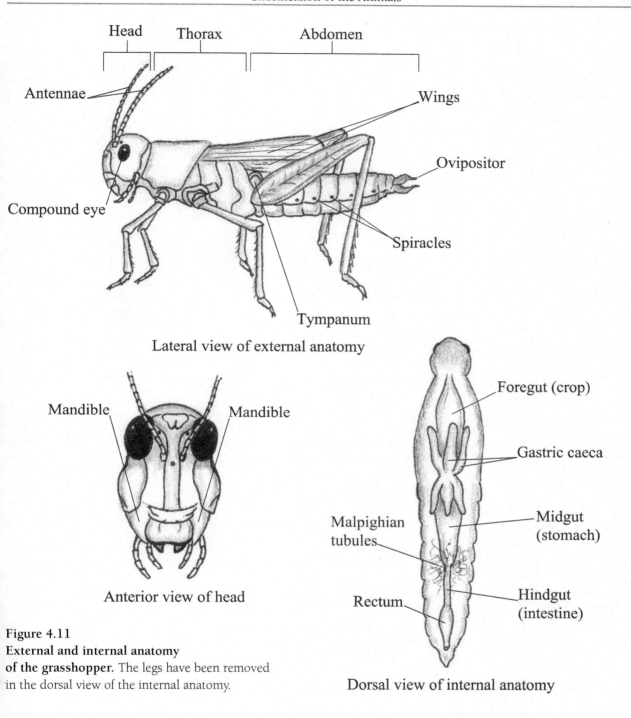

Figure 4.11
External and internal anatomy of the grasshopper. The legs have been removed in the dorsal view of the internal anatomy.

Dissect a grasshopper. First be able to identify the following external structures (Figure 4.11): head, thorax, abdomen, mandible, antennae, compound eyes, tympanum (functions as an ear), spiracles, and, for females, the ovipositor. Spiracles lead to tracheae inside the grasshopper which serve in gas exchange and thermoregulation. Most spiracles are located along the lateral abdomen. The ovipositor is used to inject eggs into the ground. For female bees and wasps, the ovipositor can be also modified for stinging.

Take off the back covering to the thorax and abdomen of the grasshopper. Slip a pair of scissors into the lateral thorax just behind the head. Cut down the side of the grasshopper. Repeat on the other side. Gently remove the back covering. The abdomen may be covered by gonads which need removal. Be able to identify the crop, stomach, gastric caeca, intestine, and malpighian tubules which create a fuzzy texture to the lower intestine. The crop has storage function. Digestion occurs in the stomach and gastric caeca. The gastric caeca have endosymbiotic bacteria and protozoans to facilitate digestion of plant material by the grasshopper. The intestine has assimilatory, osmoregulatory, and excretory function with aid of the malpighian tubules.

Phylum Echinodermata—sea stars, brittle stars, sea urchins, sea lilies, sea daisies, and sea cucumbers

The name Echinodermata comes from the Greek terms echinos for spiny and derma for skin. Hence, echinoderms are spiny-skinned animals. Most show pentaradial body symmetry where the body is divided into five parts. These eucoelomates have an open circulatory system, but no heart. Excretion and gas exchange are accomplished by diffusion. The inability of the echinoderms to osmoregulate can explain why all are marine. Many are predators, while others graze upon algae or are scavengers.

Examine the representative specimens of echinoderms.

Class Asteroidea—starfish or sea stars

The majority of starfish are hand-size in diameter, although some can be 1 m across. Examine the representative specimen(s). A few are suspension feeders, secreting mucus that traps food. Most are predators. They have two stomachs, one on top of the other. The stomach closest to the mouth can be everted into e.g., open slits in a clam, to release digestive enzymes. The more aborally (body side without the mouth) located stomach functions in digestion. Food is further digested, stored, and distributed through pyloric caeca. The madreporite is a sieve plate that functions to let water into the stone canal, ring canal, and the rest of the water vascular system. Hydraulic pressure in this system enables movement.

Dissect a starfish. Refer to Figure 4.12 and 4.13. Begin by identifying the external structures. Identify the arms, central disc, and madreporite on the aboral side, and mouth and podia on the oral side. Cut open an arm and central disc, exposing the internal structures. Be careful to cut around the madreporite and not underneath it as to keep the connection to the stone canal. Identify the pyloric caeca, gonads (under the pyloric caeca), ambulacral ridge, the soft ampullae that run along the ambulacral ridge, stomachs, stone canal and ring canal. The ampullae are the structures internal to the podia. Podia function in suction for use in movement and prey capture.

The starfish is commonly used to study the early stages of embryonic development. The zygote undergoes cell division to produce an embryo. Repeated cell divisions in the embryonic tissue are called cleavages, as the cells do not grow in size between divisions. Cleavage will form a compact ball of embryonic cells known as the morula, followed by a hollow ball of cells known as the blastula. The blastula contains a central cavity called the blastocoel. Particular cells on the blastula then divide inward creating a second cavity, called the archenteron, and two germ layers. The archenteron will develop into the digestive system. The location where cells divide inward is known as the blastopore. Examine a slide of starfish development. Locate the zygote, morula, blastula, blastocoel, gastrula, archenteron, and blastopore (Figure 4.14).

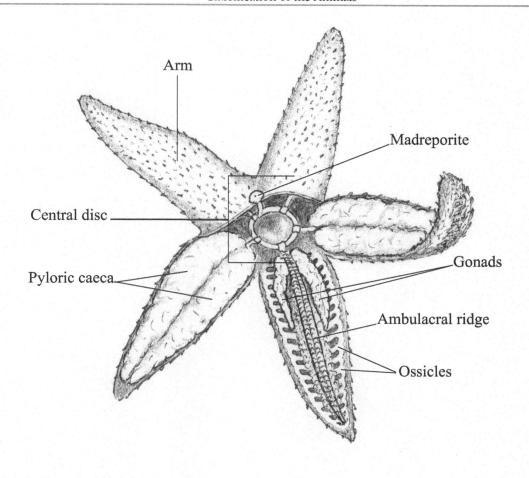
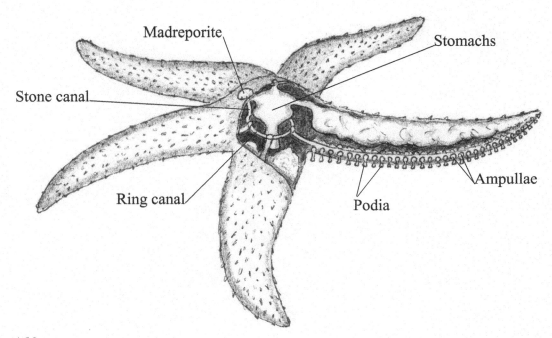

Figure 4.12
Aboral view of the external and internal anatomy of the starfish.

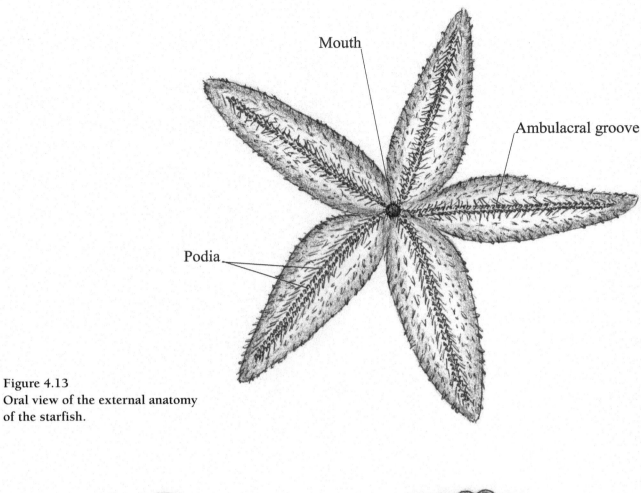

Figure 4.13
Oral view of the external anatomy of the starfish.

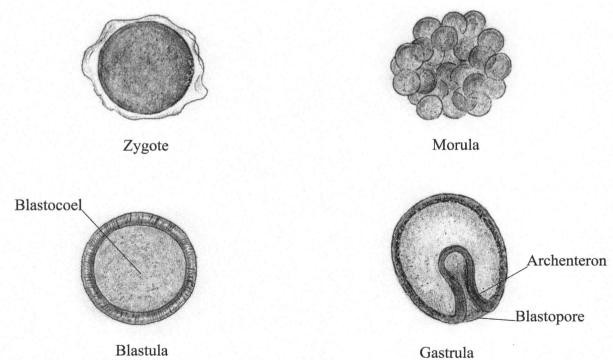

Figure 4.14
Early developmental life stages of the starfish.

Class Ophiuroidea—brittle stars

Brittle stars have a flattened central disc with reduced podia lining the arms. The body wall is ciliated except for a few areas and functions to remove debris. Most brittle stars are mobile and are predators, scavengers, deposit feeders, or filter feeders.

Class Echinoidea—sea urchins, heart urchins, and sand dollars

Echinoids are armless, and are covered with spines. In heart urchins and sand dollars, the spines are small. Many echinoids are grazers of algae.

Class Crinoidea—sea lilies

The sea lilies are an ancient group, dating back 600 million years, and being particularly common between 300 and 500 million years ago based on the fossil record. The jointed appearance of sea lilies is due to internal ossicles. Cilia close to the mouth are used in suspension feeding.

Class Holothuroidea—sea cucumbers

Sea cucumbers, unlike other echinoderms, are elongated between mouth and anus. They lack arms and are deposit or suspension feeders.

Phylum Hemichordata—acorn worms

These eucoelomate worms receive their name because of the collar that separates the trunk of the worm from the proboscis. Their appearance is that of a very elongated acorn. Acorn worms are all marine, with a closer genetic relationship to the Echinodermata than the Chordata. However, like chordates, they possess pharyngeal slits and in some species, a dorsal tubular nerve cord. Most are deposit feeders. They ingest sediment for the associated organic material.

Examine the representative specimen(s).

Phylum Chordata—tunicates, lancelets, vertebrates

All chordates are eucoelomates and possess the following structures at some point of their life cycles: a notochord, a dorsal tubular nerve cord, pharyngeal slits, and a post-anal tail. Chordates compose many of the largest animals on earth.

Examine the representative specimens of chordates.

Subphylum Cephalochordata—lancelets

Lancelets are marine filter feeders that bury themselves into the sediment, exposing only their tentacles and mouth.

Subphylum Urochordata—tunicates

All are marine. Adult tunicates are sessile, sticking to solid substrates to include the bottom of boats and sides of piers. Each animal is enclosed in a protein-carbohydrate coat called a tunic.

Subphylum Craniata—fish, amphibians, reptiles, and mammals

The craniates are characterized by a cranium that partially to fully encases the brain, and by the neural crest that is composed of embryonic cells located by the neural tube, which contribute to the development of the skeleton and other structures. All have a closed circulatory system, a liver, and a pair of kidneys.

Class Myxini—*hagfishes*

Hagfishes are marine scavengers that lack jaws, fins, and cornea and lenses in their eyes. They secrete a copious slime when threatened by predators. The cranium consists of a cartilaginous skull.

Clade Vertebrata—animals with a backbone

Class Petromyzontida—*lampreys*

Lampreys lack hinged jaws and they have a rudimentary vertebral column. They inhabit marine and freshwater habitats where they are parasitic or predatory on fish.

Clade Gnathostoma—jawed vertebrates

Class Chondrichthyes—*sharks, skates, rays, and chimaeras*

These fish have a cartilaginous skeleton and hinged jaws where the teeth are not fused to the jaws. They are marine predators.

Class Actinopterygii—*ray-finned fishes*

The ray-finned fishes compose the largest group of chordates. They have an ossified skeleton, a protective covering to the gill called the operculum, and fins supported by flexible rays of thin bones. They inhabit marine and fresh water habitats and show a range of diets.

Clade Tetrapoda—limbed vertebrates

Class Amphibia—*frogs, toads, salamanders, and caecilians*

Amphibians are tetrapods as adults, and their development typically involves metamorphosis from tadpoles with gills to the adult stage having lungs. Adults of some amphibians retain gills. Retention of juvenile characteristics is called neoteny. Amphibians are predators as adults.

Clade Amniota—tetrapods with extra-embryonic membranes

Class Reptilia—*turtles, lizards, snakes, crocodilians, tuataras, and birds*

Reptiles possess an amniotic egg with protective membranes that surround the embryo. This feature enables them to reproduce on land. Most are ectotherms, absorbing external heat as their primary heat source. Behaviors like basking in the sun, and larger body sizes of some, e.g., crocodilians, facilitate maintenance of relatively warm body temperatures. However, birds are endotherms, that is, they produce body heat from within. More precisely, they are homeotherms, maintaining a relatively high and constant body temperature.

Class Mammalia—*mammals*

Mammals have hair and mammary glands that produce milk for their offspring. They are also amniotes, although in most extant species, extra-embryonic membranes are incorporated into the mother's placenta. Like birds, they are endothermic and some may exhibit lowered body temperature during periods of torpor or hibernation. The teeth of mammals are differentiated, allowing for more complex and specialized diets. Mammals have become the dominant large animals on earth today.

EXERCISE
Classification of the Animals

Name_____ Date_____

Domain Eucarya—Kingdom Animalia

Use your lecture text in helping to answer the questions below.

1. What characteristics do protostomes and deuterostomes show? Which phyla are protostomes and which phyla are deuterostomes?

2. Contrast acoelomate, pseudocoelomate, and eucoelomate body plans. List the phyla that show each of the body plans. What are the advantages of having a body cavity like a coelom?

3. What is the function of the sponge spicule? What are choanocytes?

4. Sponges are relatively simple in structure. Why have the sponges not been replaced by more complex animals during the course of evolution?

5. What is the most commonly seen life stage of the Class Hydrozoa? And the Class Scyphozoa?

6. What are the selective advantages of a gut having both mouth and anus?

7. What is the adaptive value of excreting uric acid instead of ammonia for terrestrial animals?

8. Which type of circulatory system could be expected to be more efficient at transporting oxygen to body tissues, an open circulatory system or a closed circulatory system? Explain.

9. *Ascaris* is relatively simple in structure compared to the earthworm. Does this mean that *Ascaris* evolved first? Explain.

10. Insects tend to be smaller than many animals. What factors limit the sizes of insects? Consider morphological and ecological factors.

11. Compare the number of antennae, leg pairs per segment, and major body parts among the spider, crayfish, millipede, centipede, and insect.

	Number of Antennal Pairs	Leg Pairs	Major Body Parts
Spider			
Crayfish			
Millipede			
Centipede			
Insect			

12. Insects are characterized as having an open circulatory system. Yet, many are very active and can fly. What is the adaptive advantage of an open circulatory system?

13. What are the selective advantages to being pentaradial for adult echinoderms?

14. What distinguishes the Myxini from the Petromyzontida?

15. Both fish and lizards tend to be ectotherms, yet fish only have what is known as a single circulatory system, where blood passes once to the heart with each loop of the body circuit. Lizards have a double circulatory system where blood passes through the heart twice per body circuit. During the first pass through the heart, the blood is pumped to the lungs to be oxygenated. During the second pass, oxygenated blood returning to the heart is pumped to the tissues. A double-circulatory system supports a higher metabolism. Why would higher metabolic rate be selected for in a lizard, a tetrapod, in comparison to a fish?

16. Birds used to be listed in their own class, the Class Aves. Why are they now reclassified as reptiles?

5 Dissection of the Frog

Purpose

Dissection of the frog should provide a basic understanding of vertebrate structure. Should dissection of a frog and pig be included in your lab activities, you can contrast the structures of each. Both are tetrapod vertebrates, where the frog is a poikilotherm (body temperature varies with the environment), while the pig is a homeotherm (physiologically maintains a constant body temperature). Organ systems explored include the digestive, gas exchange, circulatory, urogenital, and reproductive systems.

Materials

Double-injected bullfrog, dissection tray, dissection kit that includes a scalpel and pair of dissection scissors, and protective gloves.

Procedure

Explore the external and internal anatomy of the frog. Be able to identify the parts of the frog listed below. Answer the questions on the data sheet.

External Anatomy

The eye has three eyelids: the upper and lower eyelids, and the movable nictitating membrane which can cover the cornea. Behind each eye is a tympanum (Figure 5.1). This flat eardrum vibrates with sound waves, causing a small white bone (stapes) attached inside to vibrate, propagating the mechanical energy onward to allow the frog to hear. The tympanums of males are much larger than those of females. External nares are visible dorsally to the mouth.

The trunk of the frog has pectoral (arm) and pelvic (leg) appendages. The pectoral foot has four distinct toes and an underdeveloped medial toe. This toe has a swollen pad if a male. Each webbed pelvic appendage has five distinct toes. An underdeveloped medial toe may also be seen. Between the pelvic legs and positioned more dorsally, is the cloacal opening. The cloaca is the posterior ending common to the digestive, reproductive, and excretory systems.

Internal Anatomy

Consult Figure 5.2 to see how to open the frog. Begin by cutting into the skin on the ventral side and just above where the lower legs meet. Proceed to cut to the lower tip of the mouth. Next make cuts between the arms and legs as shown. Carefully follow these cuts more deeply through the muscle layers. A tough sternum will be encountered as you cut through the chest. Continue the cuts between the appendages through the sides of the body

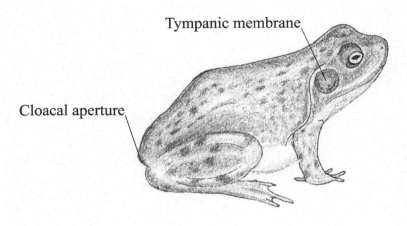

Figure 5.1
External structures of the frog.

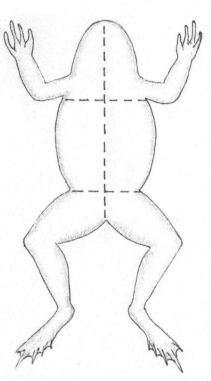

Figure 5.2
Incisions for frog dissection.

so that the skin and muscle can easily be pulled away from the internal organs. Be careful as to not cut the underlying organs.

Once the body cavity is open, you will see red and blue blood vessels. The colors are due to latex. Red latex is injected into the arteries, while blue latex is injected into the veins. Latex injection is not perfect and often colors are inaccurately injected. The organs may inadvertently be injected. Hence, be careful in identification of blood vessels. Organs and major blood vessels are held loosely in place by mesentery. This thin and transparent connective tissue will need to be cut when gently moving organs, to locate more dorsally-positioned structures.

Try to leave structures intact to observe connections.

A blue ventral abdominal vein will need to be removed once you open up your frog. If you have a gravid female frog, the ovaries and oviducts may complicate dissection. If needed, remove the ovaries and oviduct running on the left side of the frog. This will be your right side as you view the frog.

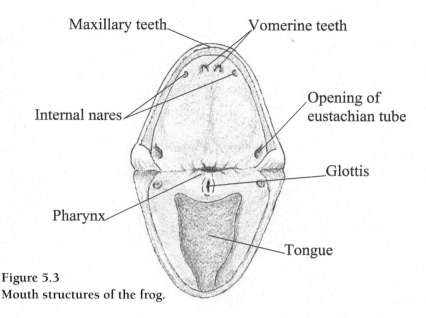

Figure 5.3
Mouth structures of the frog.

However, first read through this lab and familiarize yourself with the internal body as to not accidentally remove unintended structures.

1. The Digestive System

    The digestive system begins with the mouth (Figure 5.3). Open the mouth. You will need to use a pair of scissors to cut through the jaw bone on both sides of the mouth. Locate the maxillary teeth, internal nares, vomerine teeth, tongue, and glottis. Posterior to the mouth is the pharynx, where the glottis and Eustachian tubes are located. The esophagus follows the pharynx.

    Proceed to examine the body cavity (Figure 5.4). Follow the remaining digestive system in sequence. Locate the esophagus that runs dorsally to the heart and connects to the large stomach. Food passes through the stomach into the duodenum, which is the first region of the small intestine. The duodenum largely functions in digestion. Blood vessels are few. Running alongside and emptying into the duodenum, the pancreas contributes bicarbonate and digestive enzymes. The pancreas resembles a thin elongated flap having the same coloration as the adjacent duodenum.

    Among other functions, the multi-lobed liver produces bile, which is stored in the bulb-like gallbladder. Bile is released into the duodenum, where it serves to emulsify fat. The gallbladder is located just under the right lobe of the liver. The duodenum transitions into the jejunoileum, where digestion continues and most of the assimilation of food occurs. Note the many blood vessels connecting to the jenunoileum. Following the small intestine is the large intestine, which functions in osmoregulation. It terminates into the cloaca.

    A blackish to bluish round spleen is visible just underneath the jejunoileum. The spleen has leukocytes to detect antigen-bearing substances in the blood.

2. The Gas Exchange System

    The adult frog has three methods of gas exchange. The first two means are cutaneous gas exchange and buccopharyngeal gas exchange which involve diffusion through the skin and mouth, respectively. The third means is by lungs. Gases entering the mouth pass into the glottis and

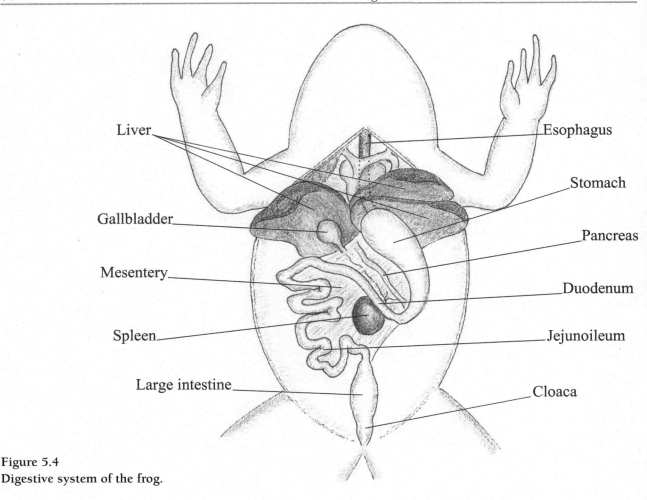

Figure 5.4
Digestive system of the frog.

on to the lungs. Locate the larynx and two lungs (Figure 5.5). Two bronchi branch from the larynx, with each entering a lung. Leave the bronchi intact until you have located all other structures in the upper thoracic cavity.

3. The Circulatory System

The heart and major arteries are illustrated in Figure 5.6. Locate the three chambered heart. Differentiate the ventricle and two atria. Oxygenated blood is pumped from the left atrium to the ventricle and then through the conus arteriosus. This vessel separates the atria and branches into a right and left truncus arteriosus. These arteries quickly branch into vessels supplying the head and lower body. Coming dorsally from each truncus arteriosus is an aortic arch. The two aortic arches transport blood to the lower body. A subclavian artery branches from an aortic arch to pass into an arm. Gentle shifting of the stomach from the left to more right side of the body will expose where the more dorsally positioned aortic arches fuse. From this union, the celiacomesenteric artery and dorsal aorta emerge. The celiacomesenteric artery transports blood ventrally to the digestive system. The dorsal aorta runs just ventral to the vertebrae and between the kidneys before branching into the common iliac arteries that travel to each leg.

Figure 5.7 illustrates the major veins of the frog. From each leg, blood will collect in the pelvic (or iliac) veins. Blood draining from each kidney is transported via renal veins to the posterior vena cava, which runs

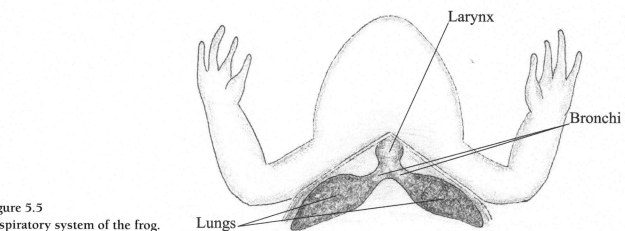

Figure 5.5
Respiratory system of the frog.

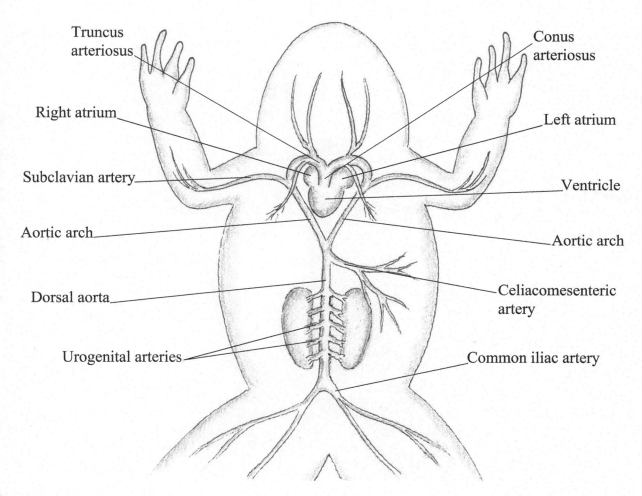

Figure 5.6
Heart and major arteries of the frog, ventral view.

through the liver to the dorsal side of the right atrium. The subclavian veins emerging from each arm and veins from the head merge on both the right and left sides of the frog to form the anterior vena cava. The anterior and posterior vena cavae join at the sinus venosus before entering the right atrium.

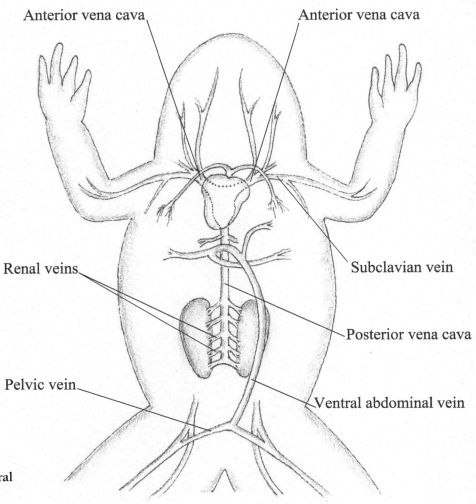

Figure 5.7
Major veins of the frog, ventral view.

4. The Urogenital System

    Refer to Figure 5.8 to help locate the structures. Paired, flat, and elongated kidneys lie on the dorsal side of the abdominal cavity. Urine flows through the ureters and collects in the bag-like urinary bladder. From here, the urine flows into the cloaca and out through the cloacal aperture.

    In a gravid female, the pepper-colored ovaries cannot be missed. Less developed ovaries will appear more yellow than black in color. Eggs released from the ovaries are transported with the aid of cilia in the body cavity to the ostium tubae located underneath the lungs. The eggs enter the ostium tubae, pass through the tubular and whitish oviduct, the membranous ovisac, and cloaca before being released out of the cloacal aperture.

    In the male, a yellowish testis can be found on the ventral side of each kidney. A testis is approximately the same size as the spleen. Sperm from the testes run through the ureters and out the cloaca.

5. The Nervous System

    The brain (Figure 5.9) is enclosed by a hard cranium. The brain can be exposed by carefully chipping and cutting away the dorsal side of the cranium (i.e., removing the nasal and frontoparietal bones). More accessible structures of the nervous system include the whitish brachial nerves that travel in each arm, and the sciatic nerves that run in each

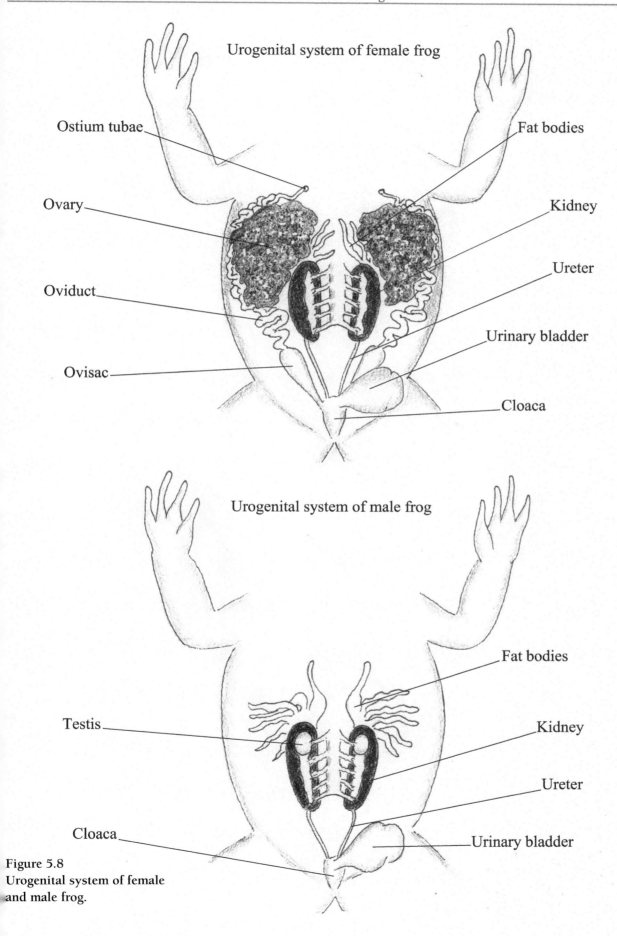

Figure 5.8
Urogenital system of female and male frog.

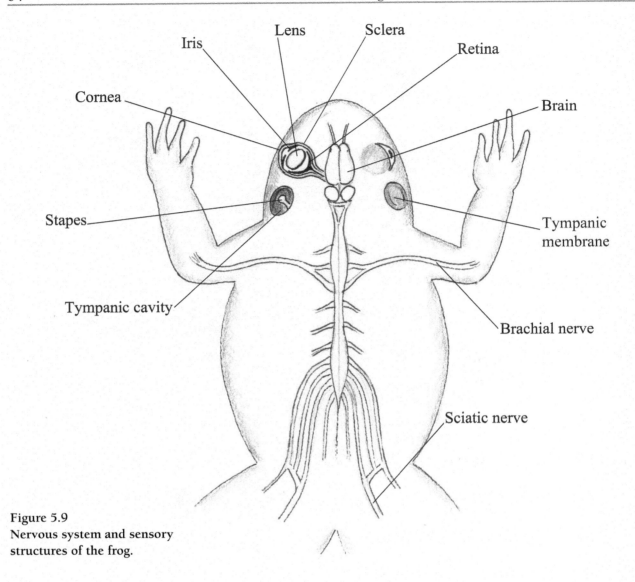

Figure 5.9
Nervous system and sensory structures of the frog.

leg. Each sciatic nerve comes from the fusion of spinal nerves clearly visible beneath each kidney. The nerves are whitish in color because they are myelinated.

Each eye is covered by a transparent cornea in front. The cornea is bordered by the white tough sclera. The whitish lens and surrounding iris are visible just inside the eye. The lens is transparent in a living frog. The retina is a blackish membrane in the back of the inner eye. Photoreceptors are concentrated in the retina.

The ear consists of a tympanic membrane to which the white and slender stapes is attached. Removal of the tympanic membrane yields the tympanic cavity.

EXERCISE
Dissection of the Frog

Name_____ Date_____

1. Few blood vessels are seen around the duodenum, unlike the jejunoileum. Why the difference?

2. Fish and frogs are both poikilotherms. Fish show single blood circulation, where blood passes through the heart once with each circuit through the body. The fish heart pumps blood through the gills. The blood then passes through the tissues of the body and back to the heart. In contrast, frogs show double blood circulation, where the blood passes twice through the heart on its cycle through the body. The right side of the heart collects oxygen-poor blood from the body and pumps it to the lungs for oxygenation. The left side of the heart then collects the oxygenated blood and pumps it to the body. Explain the adaptive advantage of the double circulation in frogs as compared to the single circulation in fishes.

3. Assume a tumor originates in the kidney of the frog and metastasizes. Why might the liver be a likely spot for the cancer to spread?

4. What function does myelin have? What is the advantage of spinal nerves and large peripheral nerves being myelinated?

6 Dissection of the Fetal Pig

Purpose

Among animals available for dissection, pigs are most like humans with regard to basic anatomy. They provide a good contrast to the frog. Like humans and unlike the frog, the heart and lungs are separated from the abdominal organs by a thick muscle known as the diaphragm. The heart of the pig has four chambers in comparison to the three in the frog, and the intestinal tract is relatively longer than that of the frog. The fetal pig has yet to be born and thus shows some fetal characteristics such as the umbilical cord.

Materials

Double-injected pig, dissection tray, dissection kit that includes a scalpel and pair of dissection scissors, and protective gloves.

Procedures

Explore the external and internal anatomy of the pig. Be able to identify the parts of the pig listed below. Answer the questions on the data sheet.

External Anatomy

Begin with the head. Two eyelids cover the transparent cornea of the eye. Refer to Figure 6.1 to locate the external nares.

Placing the pig on its dorsal side, locate the umbilical cord and urogenital opening. You should be able to identify the sex of the pig.

Internal Anatomy

Note the incision lines shown in Figure 6.1. Begin by cutting through the skin. Unlike the frog, the skin of the pig attaches more firmly to the underlying muscle. Next, cut into the muscle, being careful as to not damage structures underneath. Open the body cavity. A few extra snips to loosen the mesentery may be needed.

Once the body cavity is open, you will see red and blue blood vessels. The colors are due to latex. Red latex is injected into the arteries, while blue latex is injected into the veins. Latex injection is not perfect and often colors are not accurately injected. The organs may inadvertently be injected. Hence, be careful in identification of blood vessels. Organs and major blood vessels are held loosely in place by mesentery. This thin and transparent connective tissue will need to be cut when gently moving organs, while locating more dorsally-positioned structures.

Try to leave structures intact to observe connections. The umbilical cord and its connections may have to be cut as to move the cord to one side.

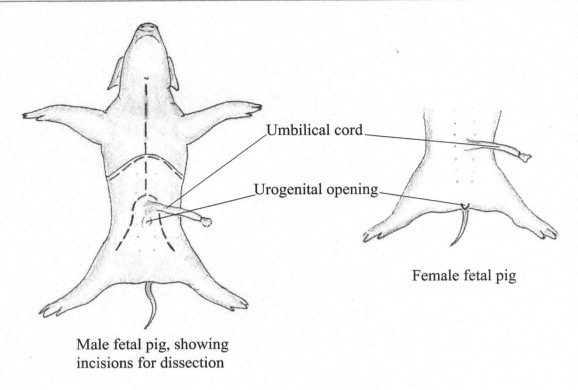

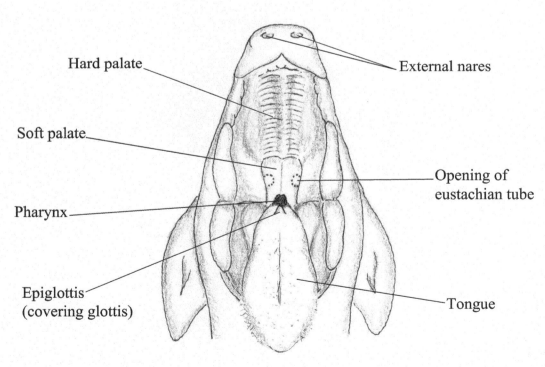

Figure 6.1
External structures of the fetal pig, top figure;
Oral cavity of the fetal pig, bottom figure.

1. The Digestive System

Begin by opening the mouth. You will need to crack open the jaws using a pair of scissors in the corners of the mouth. Continue to open the mouth so to expose the epiglottis in the back of the pharynx. The epiglottis is a covering to the glottis which leads to the lungs. Locate the tongue, the ridged hard palate, the soft palate, and openings to the Eustachian tubes. These tubes function to balance the pressure between the middle ear and atmosphere.

Use Figure 6.2 as a guide to the lower digestive organs. The esophagus is a soft white tube that lies just ventral to the similar-colored abdominal aorta found attached to the dorsal wall. The esophagus connects to the large

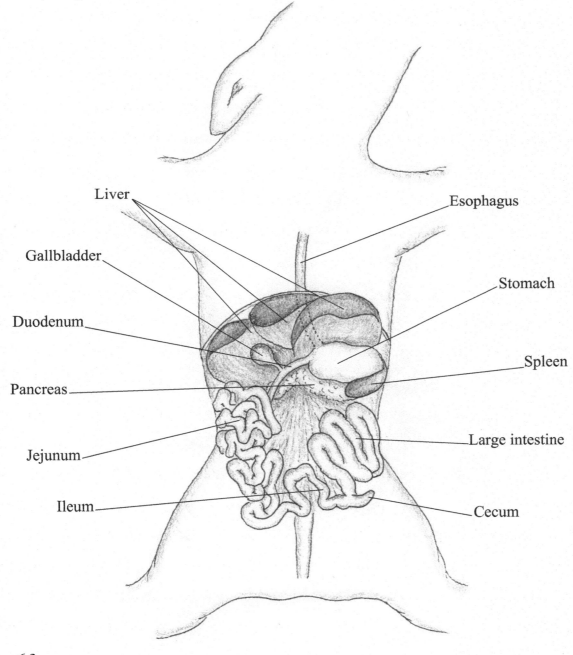

Figure 6.2
Digestive system/abdominal viscera of the fetal pig.

stomach. Acids and enzymes in the stomach participate in some digestion, plus function to kill many microbes that have been ingested. Stomach contents pass to the small intestine. The small intestine consists of the duodenum, jejunum, and ileum. The duodenum has the major function of digestion. Note the few blood vessels around the duodenum. Bile, produced in the liver and stored in the gall bladder, is passed to the duodenum via the bile duct. The gall bladder is fused to the lower right lobe of, and is usually of a slightly different color than, the liver. The pancreas is light in color and has a granular appearance. Its digestive enzymes and bicarbonates are passed to the duodenum. Digestion continues in the jejunum and ileum. It is in these regions where most food absorption occurs. The distinction between the jejunum and ileum is not clear. Both function in assimilation of nutrients. The ileum assimilates residual material, to include vitamin B_{12} and bile salt, plus has abundant lymphoid nodules.

The large intestine is thicker and usually more greenish in color that the jejunum and ileum. Lifting up the large intestine on the lower left side of the pig exposes the cecum, which is a blind pouch that emerges from the connection between the ileum and large intestine. In the pig, it contains bacteria and protozoans that aid in the digestion of plant material. The large intestine largely has an osmoregulatory role. Fecal material forms in the posterior end of the large intestine (i.e., rectum) and is egested through the anal canal.

2. The Gas Exchange System

With inhalation, air is passed into the mouth, to the pharynx, the glottis, larynx, and trachea. The trachea branches into two bronchi, each of which penetrates a pleural sac that provides the external covering to a lung. Each bronchus further branches into secondary bronchi that branch into tertiary bronchi. The bronchi continue to branch into smaller tubules, the bronchioles, before terminating into bulb-like alveoli. Each alveolus is covered by a dense network of capillaries. Gas exchange occurs between the alveolus and the capillaries.

Review the structure of the mouth. Refer to Figure 6.3 for guidance. In the dissected throat, locate the hard, round larynx. The trachea is distinguished by the cartilaginous rings that compose it. The glandular thyroid surrounds the trachea. Before proceeding to find the bronchi, locate the various parts of the heart and attached blood vessels. Locate the light-colored lungs. Can you see a difference in the number of lobes composing each?

3. The Circulatory System

Figure 6.4 illustrates the heart and major arteries. The thymus (Figure 6.3) covers the anterior upper half of the heart. This glandular structure will need to be removed to view the atria. Locate each atrium and ventricle. Carefully remove any soft tissue lying on the anterior-most section of the heart to expose major blood vessels. The pulmonary trunk is a thick artery that emerges from the right ventricle. More dorsal to this artery is the aortic arch, which transports oxygenated blood pumped by the left ventricle to the body. Branches from the aortic arch include the internal and external carotid arter-

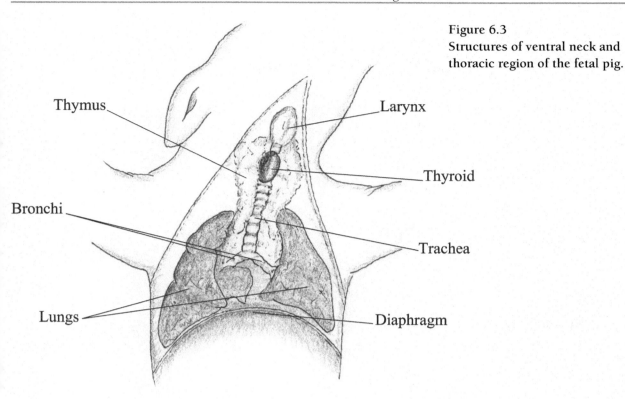

Figure 6.3
Structures of ventral neck and thoracic region of the fetal pig.

ies that run to the head, and thick abdominal aorta that transports blood to the lower body. The abdominal aorta is a light-colored vessel which attaches to the dorsal wall of the abdominal cavity. It runs just ventral to the spinal column. Careful cutting of mesentery and the diaphragm, and pushing aside of abdominal organs (to include a kidney) exposes the aorta. It branches into the external iliac arteries, which travel to the legs.

Venous blood from the legs collects in the external iliac veins (Figure 6.5). They fuse to form the posterior vena cava. The hepatic portal vein collects venous and nutrient enriched blood from the intestines, passing the blood to the liver. The liver has many functions including carbohydrate and lipid metabolism, deamination, and the production of bile. The posterior vena cava passes through the liver, further fusing with smaller veins. The thick posterior vena cava can be seen between the liver and heart. The posterior vena cava then enters the right atrium on the dorsal side of the heart.

The internal and external jugular veins run in parallel, but just more externally, to the internal and external carotid arteries. The jugulars will join with other upper thoracic veins to form the anterior vena cava which fuses with the posterior vena cava on entry into the right atrium.

4. The Urogenital System

Locate the parts of the excretory system (Figure 6.6). The kidneys are round, thick, elongated organs lying on the lower section of the dorsal wall of the abdominal cavity. At the anterior end of each kidney, you may see the membranous adrenal gland. A ureter runs from each kidney to the urinary bladder.

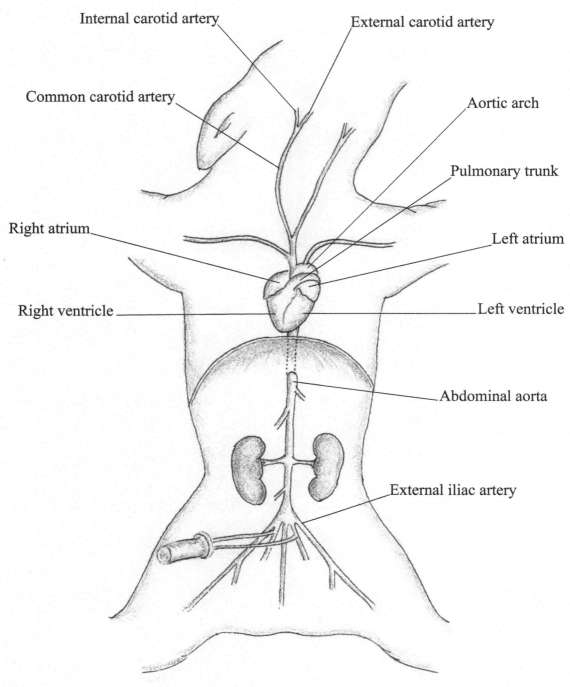

Figure 6.4
Heart and major arteries of the fetal pig.

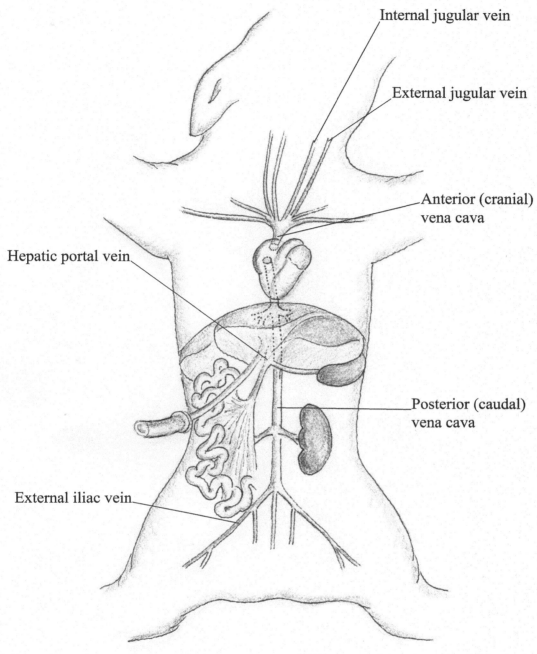

Figure 6.5
Major veins of the fetal pig.

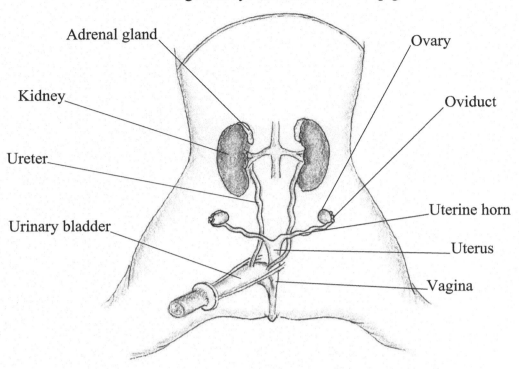

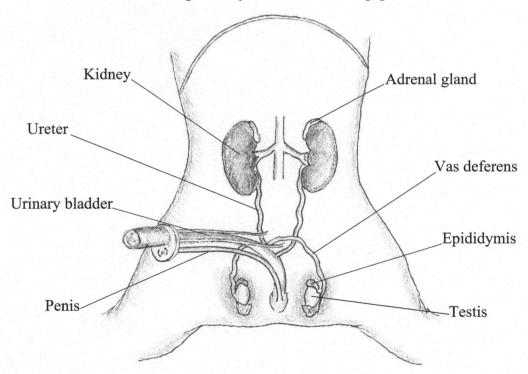

Figure 6.6
Urogenital system of female and male fetal pig.

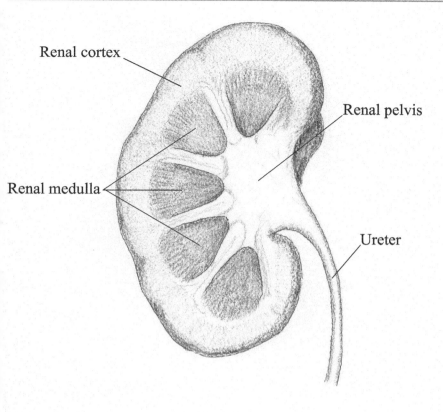

Figure 6.7
Mammalian kidney, coronal section.

Figure 6.7 illustrates a coronal sectioning of the mammalian kidney to show the internal structures. If a coronal section of an adult pig's or sheep's kidney is available, identify the parts.

Sex of a fetal pig can be determined by the location of the urogenital opening (Figure 6.1). Refer to Figure 6.6 for the female and male reproductive organs. The small round ovaries are located below the kidneys. The two oviducts running from each ovary connect to the uterus. The vagina follows the uterus, separated by the cervix.

The testes can be found embedded in the abdominal wall. The epididymis lies on the anterior end of each testis. The highly coiled epididymis proceeds to the vas deferens, which fuses with the vas deferens from the other testis to form the urethra, which runs through the penis.

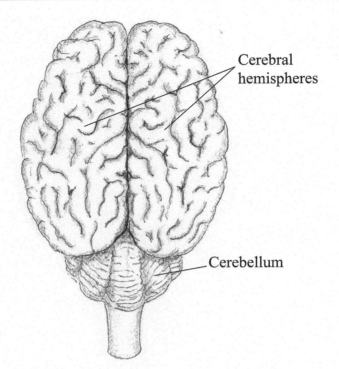

Dorsal view

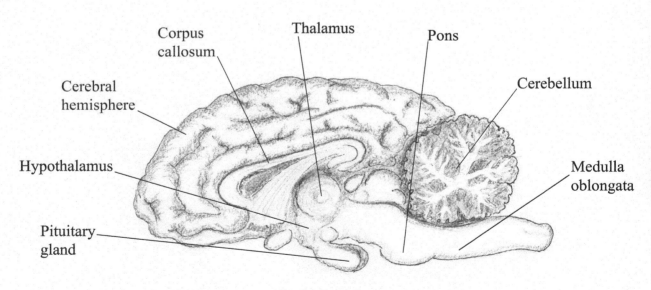

Sagittal section

Figure 6.8
Sheep's brain.

5. The Nervous System

Figure 6.8 illustrates the mammalian brain taken from dorsal and mid-sagittal views. If available, examine the brain of an adult pig or sheep. Identify the various parts. Consult your lecture text to learn the functions of each part.

The mammalian eye is illustrated by Figure 6.9. Identify the parts and know their functions. You can carefully remove the eye of the fetal pig to locate the parts. Note how small the lens of the fetal pig's eye is compared to that of the frog.

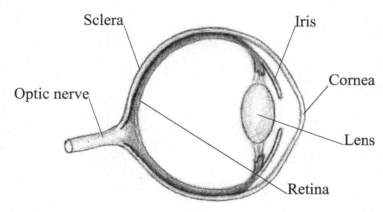

Figure 6.9
Mammalian eye, sagittal section.

EXERCISE
Dissection of the Fetal Pig

Name_____ Date_____

1. The homeothermic mammal has a longer intestinal tract relative to its body size than a poikilothermic frog. Why the difference? Explain.

2. Why do some sections of the gastrointestinal tract lack blood vessels (i.e., esophagus, stomach, and duodenum), while others are profusely transected by blood vessels (i.e., jejunum and ileum)?

3. Provide the function(s) of lymphoid nodules.

4. Compare the costs and benefits of the four-chambered heart of mammals to the three-chambered heart of lizards.

5. What is meant by osmoregulation? What organs are involved in osmoregulation?

6. Where in the kidney does the cortical nephron largely lie? And the juxtamedullary nephron? Lizards only have cortical nephrons. A juxtamedullary nephron is more efficient at extracting water from urine filtrate. Explain the presence of juxtamedullary nephrons in a mammal and the absence of them in a lizard.

7 Animal Tissues

Purpose

Histology is the study of tissues. A tissue is a group of morphologically similar cells that interact to perform specific functions. As the basic material from which an organism is constructed, tissues represent the simplest level of body organization.

Organs are composed of tissues that perform a particular function or functions. Organs include the heart and kidney. Organ systems, such as the circulatory and excretory systems, represent a higher type of functional organization.

In this exercise, you will examine tissues of animals using prepared slides. Know the functions of each of these tissues. To explore functions and learn more about tissues, consult your textbook. The four major types of tissues found in vertebrate animals (fish, amphibians, reptiles, birds and mammals) are epithelial, connective, muscular, and nervous.

Epithelial Tissue

Epithelial tissue lines body surfaces and cavities. It is bounded by free (exposed surface) on one side, and connective tissue on the other. Often, epithelial cells are modified with structures like cilia or microvilli. Epithelial tissues are composed of many tightly packed cells with very little matrix (acellular material) between cells. They function in protection, absorption, secretion, or excretion. A range of cell types compose epithelial tissue. The shape of the epithelial tissue and layering of cells have advantage in function.

See Figure 7.1 for the basic shapes and arrangement of epithelial tissue. Observe slides of simple cuboidal epithelium, stratified squamous epithelium, and pseudostratified ciliated columnar epithelium. Refer to Figure 7.2 to help locate each cell shape and arrangement.

Connective Tissue

Connective tissue is characterized by living cells embedded in a matrix. The matrix may be a solid, semi-solid, or a liquid. For bone, this matrix is the calcium deposit; for blood it is the plasma. Connective tissues are widespread throughout the body and often function to support or hold parts of the organism together. Major types of connective tissues follow below.

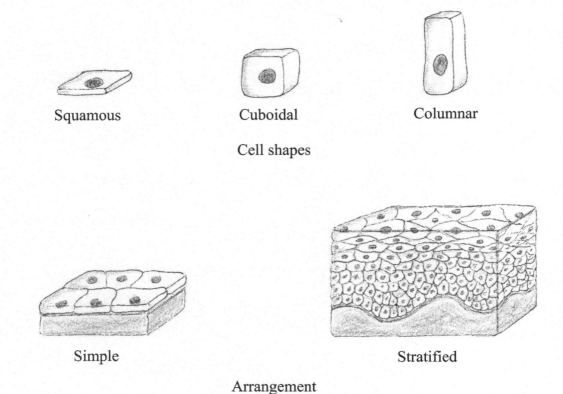

Figure 7.1
Epithelial tissue.

1. Bone

Bone functions in support, protection of soft tissues, and as a calcium reserve. A bone may contain both an outer compact layer (compact bone) and an inner spongy component (spongy bone). Compact bone is dense, while spongy bone has osseous plates arranged as a lattice. Compact bone is composed of annular units called osteons that look like the annual rings of a tree trunk in cross section. A central canal containing a blood vessel is found in each osteon. This central canal is surrounded by concentric layers of a hard apatite matrix composed of calcium phosphate, calcium hydroxide, and calcium carbonate. The layers are called lamellae. Lacunae are pits found in the lamellae that contain osteocytes (bone cells) which function in maintaining bone tissue and mechanosensation. Notice how widely separated the cells are. The compact bone of a typical long bone is surrounded by an outer fibrous layer, the periosteum, and an inner delicate membrane known as the endosteum. Bone marrow, found in the center, contains large clear adipose (fat) cells and hemopoietic (blood forming) cells.

Simple cuboidal epithelium

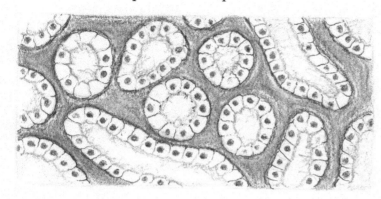

Figure 7.2
Types of epithelial tissue.

Stratified squamous epithelium

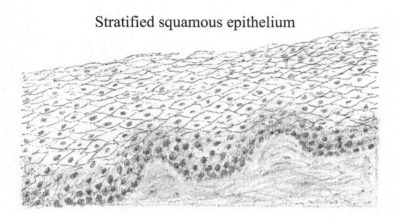

Pseudostratified ciliated columnar epithelium

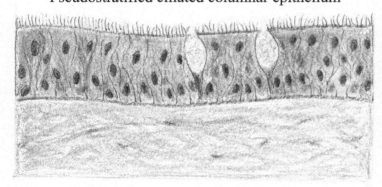

Examine the slide of ground bone (Figure 7.3) which shows the result of appositional bone growth of compact bone around a blood vessel. Be able to identify the parts.

Learn the major bones of the human body (Figure 7.4) using an actual skeleton or skeletal model. The skeletal system is most basically divided into the axial skeleton which includes the skull, vertebral column, ribs, and sternum, and the appendicular skeleton which includes the pectoral girdle and pelvic girdle, as well as the bones of the limbs.

If an actual skeleton or an anatomically correct model is available, determine gender. Consult your text or appropriate sources to learn about sexually dimorphic differences between the female and male skeleton. The pelvic ratio can also be used in gender determination. The ratio is computed from the dividend using the equation,

$$\text{Pelvic ratio} = \frac{\text{(cm distance between the tips of the ischial spines)}}{\text{(cm distance between the inner surface of the public symphysis and upper inner surface of the sacrum)}}$$

The ischial spines are sharp projections visible on the lower pelvic girdle as seen from a dorsal prospective. A ratio of ≥ 1 is characteristic of females while a ratio of ≤ 0.8 is characteristic of males.

Compare the human skeleton to those of other vertebrates, and identify the homologous bones to include how, according to the animal, they are modified according to function.

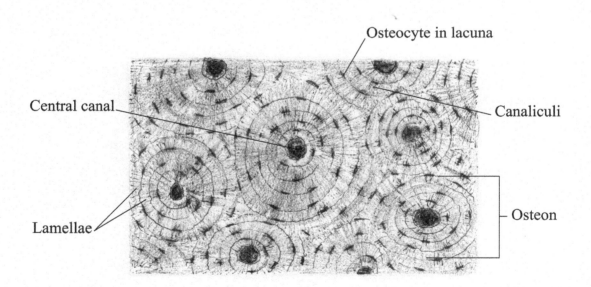

Figure 7.3
Bone tissue.

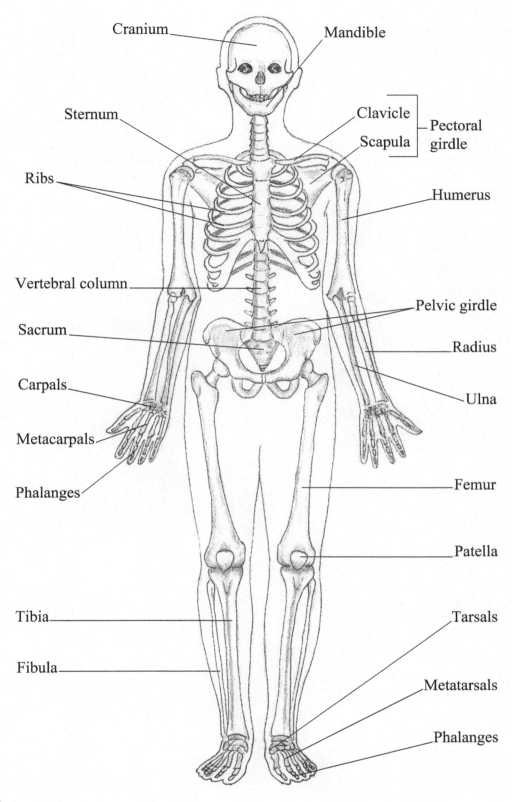

Figure 7.4
The human skeleton.

2. Cartilage

Cartilage provides flexible support, for example, between ribs and the sternum, and cushioning between long bones. The matrix is composed of collagenous fibers embedded in a rubbery substance called chondrin. This matrix is secreted by chondrocytes.

Examine a slide having hyaline cartilage. Be able to identify the chondrocytes in lacunae. Refer to Figure 7.5 for help in identifying the parts.

3. Adipose Tissue

Adipose tissue stores fat and also produces various chemicals such as the appetite suppressant, leptin. Be able to identify the adipocytes or fat cells. Refer to Figure 7.6 for assistance.

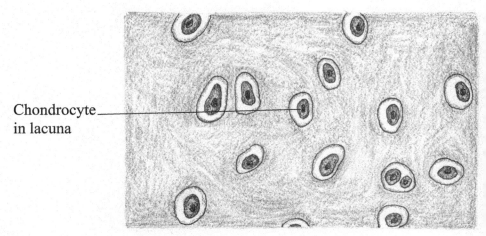

Figure 7.5
Hyaline cartilage.

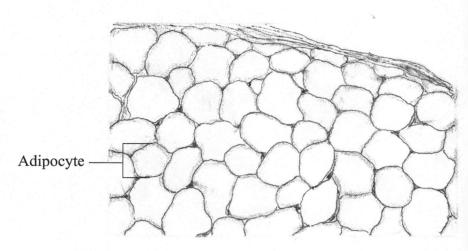

Figure 7.6
Adipose tissue.

4. Blood

Blood has many functions, including those involving blood clotting and immune responses, and the transport of nutrients, wastes, electrolytes, hormones, and gases. The fluid matrix of blood is plasma.

Observe a slide of human blood. Try to locate and identify all of the cell types in Figure 7.7. Consult your textbook regarding the functions of each of these cell types.

Muscle

Muscle is contractile and functions in movement of the body and its organs, e.g., food passage during digestion, and heart contraction. However, muscles also secrete chemicals (collectively called myokines) that have paracrine, autocrine, and endocrine effects in influencing muscle development, immune function, responses to stress, and metabolism.

Examine slide material of the three major muscle types: smooth muscle, cardiac muscle, and skeletal muscle. Refer to Figure 7.8 for assistance. Smooth muscle is widespread throughout the body, found e.g., surrounding hair follicles, and contributing to the walls of blood vessels and to the wall of structures in the digestive system and reproductive system. It is usually arranged in sheets. Cardiac muscle composes the bulk of the heart, while skeletal muscle connects by way of tendons to the skeleton. Cardiac muscle is characterized by intercalated discs or transversely oriented bands. They often are darkly stained. Skeletal and cardiac muscles are characterized by striations that run along their length. Unlike the other two muscle types, skeletal muscle is under voluntary control.

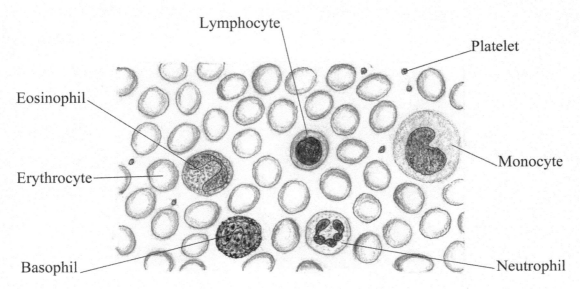

Figure 7.7
Human blood.

Figure 7.8
Muscle tissue.

Skeletal muscle

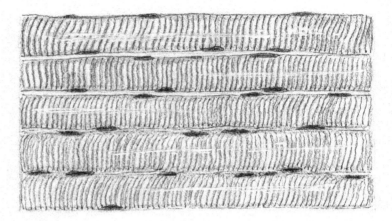

Cardiac muscle

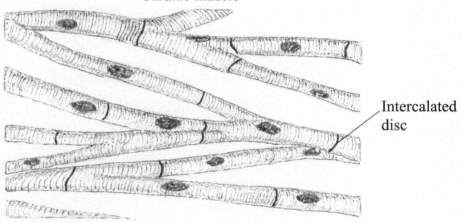

Intercalated disc

Smooth muscle

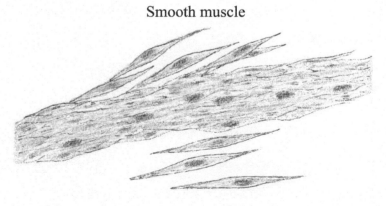

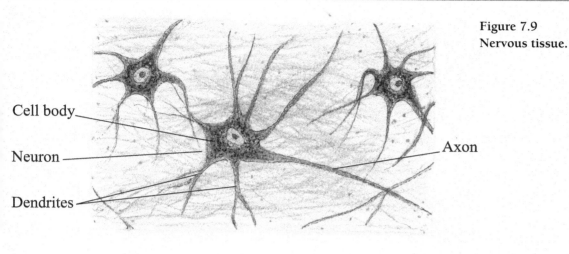

Figure 7.9
Nervous tissue.

Nervous Tissue

Nervous tissue is designed to receive stimuli and transmit electrical messages. The nerve cell is called a neuron and has three main regions: the dendrite, soma (cell body), and axon. Figure 7.9 illustrates the structure of a typical motor neuron. The dendrites are numerous short projections on one side of the neuron that are specialized to receive impulses and pass them to the soma. The cell body contains the nucleus and the majority of the cytosol. The axon is usually a single, wider, and much longer projection on the side of the neuron opposite the dendrites. It is specialized to transmit the impulse to the next neuron, muscle, or gland. Synaptic terminals are located at the tip of each axon and are responsible for secreting neurotransmitters that diffuse across the gap (synaptic cleft) to continue the message to the next structure in line. Many additional cell types are present in nervous tissue.

EXERCISE
Animal Tissues

Name_____   Date_____

1. Explain the selective value of having stratified squamous epithelial tissue composing the outer portion of the skin.

2. Explain the selective value of having simple columnar epithelial tissue lining the small intestine.

3. The ossified skeleton of ray-finned fish and tetrapods has functions in support, protection, and mobility, but also provides a source of calcium. Besides being a building block in bone and teeth, where else is calcium needed in the body?

4. Provide characteristics of female and male skeletons. Should an actual skeleton or an anatomically correct model be available, explain how you determined gender. Also provide the pelvic ratio.

   Pelvic ratio = _____

5. Construct a table that compares the homologous skeletal features of the vertebrates that you examined according to how they are modified for function.

6. Understanding the activities of many myokines, explain how good muscle health relates to the overall health of the organism.

8 The Relationship of Environmental Variations to Stomata Density

Purpose

To be able to describe the adaptive nature of stomata density and function in plants as relating to varying environmental conditions.

To develop skills in quantitative analysis and critical thinking.

Materials

Small plastic bags to store collected leaves
Permanent marker to label storage bags
Clear adhesive tape (not opaque)
Clear nail polish
Permanent marker or wax pencil
Compound light microscope (with 400 × total magnification)
Microscope slides
Boxes for storing slides
Student-generated data table

Introduction

Vascular plants use stomata to exchange gas with the above-ground environment. Stomata are microscopic pores in the plant's epidermis and occur primarily on the undersides of leaves. The open or closed state of a stomatal pore is controlled by two guard cells (Figure 8.1). When a pair of guard cells are turgid (because they are filled with water), the stoma is open. This allows carbon dioxide, CO_2, used for photosynthesis to enter the leaf but also allows water, H_2O, and oxygen, O_2, to escape. The release of water through stomata aids in transporting water and soluble materials through the vascular system. Evaporation of this water from the surface of leaves also cools the leaf, helping to prevent overheating and denaturing of photosynthetic molecules.

The opening and closing of stomata (stomata behavior) may not be the only way plants control gas exchange rates and water loss through evapotranspiration. These processes may be augmented by differences in stomata density, the number of stomata per unit area. Stomata density can differ among plant species and may even vary among leaves of a single plant exposed to different microenvironments (e.g., sun or shade). It is also possible that stomata density can vary over time for some plants, such as when leaves develop under higher CO_2 levels (Kouwenberg et al. 2003). Variations in stomata density reflect adaptations to interacting needs of sufficient CO_2 uptake for photosynthesis, the internal transport

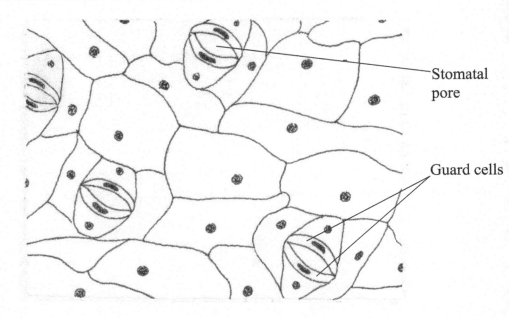

Figure 8.1
Leaf stomata among epidermal cells, 400×.

of water and other nutrients, thermoregulation, and the risk of excessive evapotranspiration which can lead to desiccation. Thus, there may be several alternative hypotheses relating environmental conditions to stomata density. As an investigator, you will need to formulate one potential hypothesis and collect leaves exposed to different microenvironments to test its predictions.

Procedures

Part 1 – Methods for Obtaining Leaves

1. Choose a location on campus (or another site—get prior approval from your instructor) where you can find green, living leaves. Given the variety of natural areas and horticultural plantings on campus, you should have no trouble, even in winter. If, however, this is a problem, some robust, deciduous leaves, such as oak leaves, may be used instead.

2. Collect a total of 20 leaves from a given plant species or variety—10 from one particular microenvironment and 10 from another (one of these may serve as your "control"). Place each group of samples in its own labeled plastic bag to bring back to the lab within 24 hours of collecting (the sooner the better). If you want to add another dimension to your investigation, collect groups of leaves from plants of different species; ensure that you have the same representative microenvironments to enable within-species and between-species comparisons. Alternatively, your classroom may be equipped with plant growth chambers that allow adjustment of light, humidity, or CO_2 levels. Students should work together to set up experimental groups (control and treatment plant groups) to investigate the potential effects of a particular environmental variable on leaf stomata density.

Part 2 – Methods for Obtaining Stomata Impressions

The following methodology is adapted from Brant and Vatnick 2004.

1. Take one of the leaves that will be used for obtaining stomata density. Determine which side of the leaf you will use to count stomata (usually the underside). Paint a thick layer of clear nail polish on this side. Allow the nail polish to dry at least several minutes. Stick a section of clear tape completely and firmly on the section of nail polish. Slowly peel the tape away from the leaf; the nail polish should stick to and peel up onto the section of tape. This is your leaf impression. Tape this leaf impression to a clean slide. Cut any overhanging tape from the sides of the slide. Use your marker or pencil to label the slide with a unique code (e.g., treatment and leaf number). Repeat for remaining leaves.

2. Design a data table in which to record stomata counts. Ideally, you should set up this table in a spreadsheet program, such as Microsoft Excel. Be sure to designate each leaf, experimental group, and species if applicable. To perform stomata counts, observe the leaf impression under 400× magnification. Choose an area reasonably representative of stomata density (or at least an area with a relatively large number of stomata). Count all stomata that are entirely visible in the field of view. Repeat two more times so that a total of three separate areas are counted on one leaf impression.

3. Average across these three counts and then divide the average by the diameter of the field of view to obtain the density of stomata for a single leaf impression. (Recall that the area of a circle $= \pi *$ radius^2. Your measurement of the area of the field of view at 400× should be less than 0.5 mm^2. Report your density as number of stomata per mm^2.) Remember, you should have a minimum of 10 leaves (i.e., independent samples or replicates) for each experimental group. (Important: although you will perform three counts per leaf impression, these counts cannot be considered independent samples—your density calculation is the independent datum.)

Part 3 – Data Analysis

If you have not already done so, it is recommended that you enter your stomata density data into a spreadsheet program. You may then use this program to calculate average stomata density per experimental group (or per experimental group separated by species), its standard deviation and standard error of the mean. These are descriptive statistics—use this information to generate a graph of your data. Be sure to plot not only the averages but also include standard error bars to show the amount of variation about each mean. You will then test your hypothesis through the use of inferential statistics, in this case a t-test. (See the lab "An Introduction to Biometry" in this manual. You may again either use a computer program or perform the calculations by hand.) Report results of your calculations and conclusion. Your conclusion should offer explanation to the results and provide critical comment as to data support of your initial hypothesis.

Suggested Activity

Write a scientific paper on this lab investigation. Be sure to follow the guidelines given by your instructor and those in the lab, "Writing a Scientific Paper."

References

Grant, B. and I. Vatnick. 2004. Environmental Correlates of Leaf Stomata Density. Teaching Issues and Experiments in Ecology, TIEE Volume 1, Ecological Society of America. (www.tiee.ecoed.net)

Kouwenberg, L. L. R., J. C. McElwain, W. M. Kurschner, F. Wagner, D. J. Beerling, F. E. Mayle, and H. Visscher. 2003. Stomatal frequency adjustment of four conifer species to historical changes in atmospheric CO_2. American Journal of Botany 90:610–619.

EXERCISE
The Relationship of Environmental Variations to Stomata Density

Name_____ Date_____

1. Explain the mechanism for stomata opening and closing. Specify the roles of ions and any plant hormones. Explain how humidity and water availability affect stomata opening and closing.

2. Given your answer to Question 1, why would plants vary stomata density rather than stomata size?

3. Explain the important role of stomata in transporting materials through a plant's (especially a tall tree's) vascular system.

4. Why might it be adaptive for stomata to occur mostly (if not entirely) on the undersides of leaves? Which plants may not show this pattern at all? Which plants have stomata only on the upper surface of a leaf?

5. Explain the adaptive advantages of the C_4 and Crassulacean Acid Metabolism (CAM) pathways. How does this influence stomata behavior? How might this influence stomata density in these plants?

6. Describe at least 4 differences in leaf morphology that would reduce water loss in a dry environment.

7. How might stomata density serve as a bioindicator for monitoring the response of plants to changes in greenhouse gas concentrations in the future? (Hint: what does your data or those of other researchers indicate for how stomata density varies with CO_2 concentrations?) How might stomata density serve as a bioindicator for estimating CO_2 concentrations in the past (paleoclimates)?

8. Explain the adaptive advantage for stomata occurring only in the sporophyte and not gametophyte life stages of mosses.

9. Ozone, O_3, is produced in the lower atmosphere from the reaction of some air pollutants and oxygen. It is highly reactive and damages living tissues. Surmise the effects of ozone on stomata density in plants of urban versus rural areas.

9 Analyzing the Structure of Plant Communities

Purpose

To be able to measure the community structure of plant communities using quadrat analysis and transect analysis.

To be able to critically examine under what circumstances each sampling technique might best be applicable.

Materials Needed

Meter sticks and string of 1 m length, or metric tape measures
Scientific calculator

Introduction

Community structure is defined in terms of the kinds, abundance, and distributions of living organisms inhabiting a community. Knowledge of community structure is valuable to the ecologist who investigates population interactions. For example, the abundance and apparent dominance of pine in forests shared by hardwood trees have prompted ecologists to investigate the reasons for this numerical prevalence. Subsequently, ecologists have discovered that pines utilize allelopathy to out-compete hardwoods (Tobiessen and Werner 1980). Here, you will be introduced to two methods frequently used to evaluate the structure of a plant community, that being quadrat analysis and transect analysis.

Quadrat analysis involves sampling the abundance and dominance of plants within a square plot of land. In contrast, transect analysis involves sampling a certain narrow distance, say 1 meter, on either side of a long linear path. A researcher might have questions of which of the two techniques to use. Choice may be dependent on what the researcher wants to understand from analysis and other considerations, such as how much variance is observed with each new measurement. For example, if the researcher desires to sample the relative abundance of plant species, the technique that provides more consistent results for the most common species may provide greater confidence in the use of the technique. By taking a few samples and doing statistical comparisons, the researcher may be able to detect which technique tends to be the most consistent. With the choice of technique made, the researcher could take additional measurements using the technique to insure a good sample size (i.e., sampling of the community).

Procedures

The following analyses are to be done in the same wooded area. Limit your analysis to that of trees having a circumference of 15 cm or greater.

Part 1—Quadrat Analysis

Select five sample sites. These sample sites should be spread over the wooded area, but attempts should otherwise be randomly selected. Limit the size of the quadrats to 10 m × 10 m, or 100 m². In each of six quadrats, identify and measure the circumference of each tree at 1 m height. Your instructor will help in identification. If a "tree" has more than one trunk, you might treat each trunk as a different individual and proceed to measure each of these individuals. Seeds are often deposited in clusters, resulting in trees that fuse together as they grow. However, multiple trunks may arise from one individual tree as suckers (lateral stems) develop at or near ground level from the main trunk. This is less often encountered among trees growing in wooded areas. In this case, how might the suckers vary in diameter from the main trunk? Include this type of tree as you record your data if the sum of all circumferences divided by 2 is roughly 15 cm or greater. Make sure to record all of the circumferences for this type of tree. Where multiple fused trunks arise from the ground, agree on a method among your classmates of how to treat the trunks, i.e., as multiple individual trees or as a single tree. Develop a data collection sheet that lists the types of tree species and the circumferences of all individuals of each of these species identified per quadrat.

Part 2—Transect Analysis

Transect analysis is commonly used when vegetation type is patchy. The transect is a narrow strip that extends through a community. Only plants along the strip are enumerated. Select six transects at random. Information from all tree trunks within 1 m distance are to be recorded as discussed in Part 1.

Part 3—Summarization of Data

Proceed to compute the relative density and dominance of each tree species per replicate measurement. Relative density is defined as the proportion of individuals which a species contributes to all individuals found in a sample. Relative dominance reflects size and is interpreted as the contribution of the area covered by tree trunks of a particular species. Relative density does not necessarily provide information of a species importance in, e.g., energy flow or nutrient cycling, within a community. A single huge specimen may contribute more to the dynamics of a community than ten very small trees of another species.

Determine the relative density of a species as,

$$= n/N.$$

where n = the number of individuals of species i and N = the number of individuals of all species recorded in sample area.

Determine the relative dominance of a species as,

$$\text{Relative dominance} = d/D,$$

where d is the dominance computed for species i and D is the sum of dominance values for all species in the sample area. The quantity d is computed as the sum of all basal areas for individuals of species i, where basal area is evaluated as,

$$= c^2/(4\pi) = c^2/12.57,$$

where c is the tree circumference at 1 m height. The units in this equation are cm^2/m^2. For a tree having multiple trunks, you will need to compute area for each of the trunks ($= c^2/(4\pi)$), and then sum all the areas to obtain a basal area measurement for the tree.

An importance value, I, can be computed for each species by summing the relative density and relative dominance for the species. The importance value should range from 0 to 2.

Part 4—Summarize the Data for Each Measurement Technique

Compute the sample means and sample standard deviations for the relative densities and relative dominances among the six samples. Standard deviation provides an estimate of variation about the mean. Formulations are provided below.

Sample mean $= \bar{x} = \Sigma x_i/n$, where x_i is a replicate value in the sample of 6 and n is the sample size (e.g., say elm had relative dominance values of 0.2, 0, 0, 0.3, 0.1, and 0. $\Sigma x_i = 0.6$ and the $n = 6$. Hence, the mean relative dominance of elm is (0.6/6) or 0.100).

Sample standard deviation $= SD = \sqrt{[\Sigma(x_i - \bar{x})^2/(n-1)]}$. The sample standard deviation for the relative dominance of elm is the square root of $[((0.2-0.1)^2 + (0-0.1)^2 + (0-0.1)^2 + (0.3-0.1)^2 + (0.1-0.1)^2 + (0-0.1)^2)/(6-1)]$ or 0.126. In the example for elm, the sample mean ± SD is 0.100 ± 0.126.

References

Tobiessen, P. and M. B. Werner. 1980. *Hardwood seedling survival under plantations of Scotch pine and red pine in central New York.* Ecology 61:25–29.

EXERCISE
Analyzing the Structure of Plant Communities

Name _____ Date _____

Part 1—Quadrat Analysis

Tree Type	Relative Density per Quadrat					Relative Dominance per Quadrat				
	1	2	3	4	5	1	2	3	4	5

Part 2—Transect Analysis

Tree Type	Relative Density per Transect					Relative Dominance per Transect				
	1	2	3	4	5	1	2	3	4	5

Part 3—Summarization of Data

Plant Type	Quadrat Analysis Mean ± Standard Deviation	
	Relative Density	Relative Dominance

| | Transect Analysis | |
| | Mean ± Standard Deviation | |
Plant Type	Relative Density	Relative Dominance

1. Compare the data derived from the quadrat and transect sampling techniques. How do the means for relative density and relative dominance compare?

2. Which technique do you believe was better at estimating the structure of the woodland community? Explain.

3. How can standard deviation be used to evaluate and choose a particular sampling technique?

10 Allelopathy

Purpose

To be able to describe what allelopathy is and how allelopathy can affect the structure of plant communities.

To develop skills in quantitative analysis and critical thinking.

Materials

Spade or small shovel
Roots and shoots of plants removed from the field
Radish seeds
Small Petri dishes (5 cm diameter)
Paper toweling and filter paper
Distilled water, 5 10-ml graduated cylinders, 5 100-ml beakers, blender

Introduction

Interspecific competition is defined as competition between different populations of organisms. Two forms of interspecific competition are exploitative (or scramble) competition, where populations directly compete for a common limiting resource, and interference (or contest) competition, where one population restricts access to a resource by another. Allelopathy provides an example of the latter. Allelopathy occurs when one competing population produces a chemical, or allelochemical, that inhibits the growth of a competitor. Antibiotics are examples of allelochemicals which are produced by various microorganisms in efforts to gain a competitive edge against other microorganisms. Many plants also show allelopathy, which is believed to be an important mechanism that regulates the ecological succession and structure of plant communities.

Plants release allelochemicals in a variety of ways. Allelochemicals may leach from dead tissues and pass directly into the soil. Living plant tissue can also secrete allelochemicals. The following exercise examines tissues of various plant species for the presence of allelochemicals using radish seeds. The radish seeds are to be soaked in a solution of macerated plant tissues and distilled water. Poor rates of germination or initial growth of seedlings may suggest the presence of allelochemicals in the plant tissues.

Procedures

Generate a question that you will address in this exercise. Record your question on the data sheet. Consider testing a hypothesis about natural conditions that might select for allelopathy. For example, would plants in a relatively new successional community tend to show allelopathy more so than plants characterizing a community nearer to a climax state? After all, there are other ways to compete.

1. Collect the shoots (leaves plus stems) and roots of various plants. Make sure to keep samples separated according to plant type and tissue type. Aster, fleabane, sunflower, walnut, day lily, and sassafras are ideal plants to sample since they are known to contain allelochemicals.

2. Keeping track of the shoots and roots, weigh out 5 g of each tissue type. Add distilled water to fill a beaker to the 100 ml mark. Blend to macerate the tissue and then filter the mixture. Use the filtrate in step 3.

3. Set up 3 to 5 Petri dishes (5 cm diameter) per type of plant and tissue, each having 10 seeds of radish on top of a disc of paper toweling or filter paper. Keep constant the number of replicates per treatment. The paper at the bottom of the Petri dish acts to retain moisture and prevents the seeds from drying out prior to germination. Add 5 ml of the filtrate to each replicate dish. With a grease pencil, record the plant type and whether the tissue was shoot or root on top of each petri dish. Follow the same procedure with the remaining tissues.

4. How should an experimental control be created? Create the proper control conditions, complete with replicates.

5. Place all of the dishes in a sunny location or under bright lights.

6. After 3 to 4 days, record the number of seeds that germinated. You can also record the length of shoot or root of the germinating seedlings. Compare results of the tissue treatments to that of the control. Use the Chi-square test to interpret results if you were asked to use only the total number of seeds germinating from all replicates of a treatment. A two-sample, one-tailed t-test could likewise be used to test for differences in mean germination or seedling growth between the tissue treatment and the control. Record which plants and tissues showed an inhibitory effect on radish seed germination or parameter of seedling growth.

EXERCISE
Allelopathy

Name_____ Date_____

Question examined in this exercise:

Treatment (identify the plant type and tissue)	Number of Seeds Germinating per Treatment (separate replicate measurements with a comma)	Total of All Seeds Germinating among All Replicates of a Treatment
_____	_____	_____
_____	_____	_____
_____	_____	_____
_____	_____	_____
_____	_____	_____
_____	_____	_____
_____	_____	_____

1. What is your conclusion concerning the initial question you addressed by your choice of plants? Also provide any statistical summary that supports your conclusion.

2. What would be the advantages of placing allelochemicals in the following locations?
    a. Shoot (leaves and stems)

    b. Root

3. What effect could allelopathy have on the species composition of plant communities? Can allelopathy act to promote plant biodiversity? Explain.

4. How could subjected plants evolve to counteract allelopathy?

5. Why evolve allelopathy and not some other form of competitive interaction?

11 Measuring the Distributional Patterns of the Tall Goldenrod and Gall-Making Host

Purpose
To be able to determine how distributional patterns of stationary organisms can be measured and to develop critical thinking skills.

Materials
Scientific calculator
Population of tall goldenrod in a local outdoor area

Introduction
Individuals in a population can be distributed in three ways: random, uniform, and clumped. Random distribution is relatively rare in natural populations. Intraspecific competition can lead to uniform distribution. Most populations tend to show clumped distribution. Protection from predation by aggregating, location of resources and mates are factors that can cause clumping. In the following exercise, the distributional patterns of the tall goldenrod (*Solidago altissima* L., Asteraceae) and the gall-making fly (*Eurosta solidaginis* (Fitch), (Diptera: Tephritidae) are to be examined.

The tall goldenrod is an abundant species found in old fields and disturbed sites throughout the Midwest. Patches of the perennial clonal herb are commonly discovered infested by the tephritid fly as recognized by the presence of one or two spheroid-shaped galls, or "ball galls" on the stem of a plant. Female flies oviposite a single egg into the terminal bud of a stem during spring (Weis and Abrahamson 1985). The emerging larva then tunnels through the bud to just beneath the apical meristem. Actions of the larva causes the initiation of a gall which continues to expand until midsummer, obstructing the flow of nutrients to the apical meristem and reducing flower and seed production. The larva feeds within the inner tissues of the gall until it enters diapause come fall. Mature adults emerge from the galls during the following spring.

Galls are conspicuous, and are pecked open by downy woodpeckers and black-capped chickadees which look for the succulent white larvae of the fly to eat (Cane and Kurczewski 1976). The larvae are also preyed upon by the parasitoid wasps, *Eurytoma obtusiventris* and *E. gigantea* (Hymenoptera: Eurytomidae)(Abrahamson et. al. 1983).

The parasitic interaction between the tall goldenrod and the tephritid fly, confounded by predatory pressures on the fly, provides an intriguing study of the interplay of factors that determine patterns of distribution. The objectives of the following exercise are to determine the distributional patterns shown by the two symbionts and to develop hypothetical explanations to the types of distributions shown.

Procedures

1. Break into groups of 3 or 4 individuals. Each group will be responsible for counting the number of tall goldenrods, and the number of ball galls in 20 quadrats.
2. In the classroom, share your information and determine the type of distribution.

The type of distribution can be determined by comparing the product of standard deviation^2/mean, where standard deviation^2 equals the sample variance in the density of the goldenrod or fly from quadrats. Sample variance is computed by,

and the sample mean is computed by,

If the product, $SD^2/\bar{x}$, is approximately 1, random distribution is assumed; if greater than 1, clumped; and if less than 1, uniform.

For an example of calculations, suppose three quadrats yielded 10, 20, and 30 goldenrods per quadrant, respectively. The variance among goldenrod is computed as,

$SD^2 = (1400 - 60^2/3)/(3 - 1) = 100$, where the sample mean = $(10 + 20 + 30)/3 = 20$. The $SD^2/\bar{x} = 5$. Thus, the distribution of the goldenrod would be evaluated as being clumped.

References

Abrahamson, W. G., Armbruster, P. O., and G. D. Maddow. 1983. Numerical relationships of the *Solidago altissima*-stem gall insect-parasitoid guild food chain. Oecologia (Berlin) 58:351–357.

Cane, J. H. and F. E. Kurczewski. 1976. Mortality factors affecting *Eurosta solidaginis* (Diptera: Tephritidae). J. New York Entomol. Soc. 84:257–282.

Weis, A. E., W. G. Abrahamson, and K. D. McCrea. 1985. Host gall size and oviposition success by the parasitoid, *Eurytoma gigantean*. Ecological Entomology 10:341–348.

EXERCISE
Measuring the Distributional Patterns of the Tall Goldenrod and Gall-Making Host

Name_____   Date_____

Statistical Summarization of Count Measurements

Tall Goldenrod
Variance	Mean	Variance/Mean

Ball Gall
Variance	Mean	Variance/Mean

1. What pattern of distribution did the tall goldenrod show? What factors may explain this type of distribution shown by the tall goldenrod?

2. What pattern of distribution did the tephritid fly show? What factors may explain this type of distribution shown by the tephritid fly?

12   A Field Study of the Reproductive Behavior of Red-Winged Blackbirds

Purpose

To be able to identify red-winged blackbirds from other bird species and be able to distinguish red-winged blackbird adult males, adult females, and young.

To be able to recognize and characterize local habitats in which red-winged blackbirds breed.

To be able to observe red-winged blackbird behavior in the field and employ a chi-square analysis to test hypotheses with categorical data.

Materials

Access to an area having red-winged blackbirds
Binoculars
Field guide for bird identification
Pencil and datasheets

Introduction

Various factors influence behavior of organisms in their natural environment. These may include microhabitat variation, food availability, competition with other organisms, predation risk, and physiological constraints. Through field observations, it is possible to examine one or more variables at a time to reveal ecological factors that help shape animal behavior. In this investigation, we will attempt to describe the reproductive behavior of red-winged blackbirds in a local wetland. Red-winged blackbirds (*Agelaius phoeniceus*) can nest in a variety of habitats that contain graminoid (grass or grass-like) vegetation. Males establish territories and compete with other males for resources related to mating; females incubate eggs and their choice of mates appears related to quality of male territory. Do background research on the natural history of the red-winged blackbird. Consider the resources needed by this species and the variety of factors in the environment that can influence reproductive behavior. You must generate one or two hypotheses relevant to the investigation. You will be responsible for individual data collection that will be used in statistical analyses.

In this exercise, you will have the opportunity to participate in a field-based ecological/behavioral study on how variations in the natural environment may differentially affect the reproductive behavior of wild animals. You and your classmates will be collecting data that may become part of a database which will be used for future analyses and education.

Procedures

Part 1—Field work

Locate an appropriate observational site such as a cattail marsh where red-winged blackbirds are known to be found.

The instructor will provide behavior descriptions that you will attempt to observe and record on the data sheet for this lab. This datasheet may be photocopied if you need additional ones for your study. Conduct a minimum of three 1-hr observations done during different days when birds are active. Look primarily for male red-winged behavior; try to ascertain the boundaries of each male's territory. Record whether you observed the behavior in woody or grassy vegetation. You may enhance your data set by recording sets of observations on individual birds (at each observation time), recording female to male ratio, replicating observations at different time periods, and replicating observations under different weather conditions. You may work in small groups of up to 3 students; each student will fill in data sheets to be turned in *individually* on the lab assignment due date; record the names of other students with whom you worked (5 points will be deducted for each additional group member beyond maximum; data sheets of unapproved groups will be considered plagiarized).

Part 2—Data analysis

You will perform a chi-square analysis, testing the null hypothesis of no difference of the observed behaviors in woody vs. grassy vegetation. Refer to the lab "Introduction to Biometry" in this manual for a review of the chi-square analysis. You may perform additional analyses on any other hypotheses you may have generated (think of other variables to which red-winged behaviors may be correlated).

The chi-square analysis is to be performed on *the behavior with the highest count* [from frequency of occurrence data] *across all sampling days* or you may need to *sum across counts of several regularly observed behaviors*. Remember, you are to compare observed behaviors in woody vs. non-woody vegetation. Based on the null hypothesis, you expect ½ of the total observations to be in woody vegetation and ½ of the total observations in non-woody vegetation.

Suggested Activity

Write a scientific paper on this lab investigation. Be sure to follow the guidelines given by your instructor and those in the lab, "Writing a Scientific Paper."

References

Stokes, D. 1979. A Guide to Bird Behavior, Vol. 1. Little, Brown, and Company, Boston, MA, USA.

Modified from: Anderson, B. J. 2005. Behavior of Common Wetland Birds: Red-Winged Blackbirds *In* pp. 275–286, Survey of Biology 1100 Laboratory Guide, 4th edition. Stipes Publishing.

EXERCISE
A Field Study of the Reproductive Behavior of Red-Winged Blackbirds

Name_____ Date_____

Questions

1. What is a territory? Why is it important to an animal?

2. Describe three general ways that animals indicate their territory. Which one/s apply to red-winged blackbirds?

3. Define the terms monogamy and polygamy. Which one applies to red-winged blackbirds?

4. Explain how you can distinguish between adult red-winged blackbird males and females.

5. Red-winged blackbirds are migratory species. Where do the red-wing blackbirds overwinter? Why do they migrate?

6. Describe the diet of red-winged blackbirds.

7. Describe an optimal habitat for a red-winged blackbird.

8. Unlike some other migratory bird species, red-winged blackbirds have not experienced significant declines. Why might this be?

9. State your hypothesis. Perform a chi-square analysis; show your work. State whether your results support your hypothesis or not.

Red-Winged Blackbird Behavior Data Sheet

Observer Name _____

Date _____

Location _____

Habitat description _____

Weather _____

Total number birds observed: males _____ females _____

Observed Behavior (*Stokes 1979*)	Sex	Frequency of Occurrence	Duration (minutes)	Vegetation Type (woody vs. grassy)
Territory: *Songspread*				
Territory: *Song-flight*				
Territory: *Bill-Tilt*				
Courtship: *Crouch*	M			
Courtship: *Sexual-chase*				
Mating: *Copulation*	M/F			
Nest-building	F			
Eggs and Young: *Tail flick*				
Eggs and Young: *Warning call*	M			
Eggs and Young: *Warning call*	F			
Eggs and Young: *Wing-flipping*	F			
Eggs and Young: *Eggs visible in nest*	N/A			
Eggs and Young: *Young visible in nest*	N/A			

Notes: _____

Red-Winged Blackbird Behavior Data Sheet

Observer Name _____

Date _____

Location _____

Habitat description _____

Weather _____

Total number birds observed: males _____ females _____

Observed Behavior (Stokes 1979)	Sex	Frequency of Occurrence	Duration (minutes)	Vegetation Type (woody vs. grassy)
Territory: *Songspread*				
Territory: *Song-flight*				
Territory: *Bill-Tilt*				
Courtship: *Crouch*	M			
Courtship: *Sexual-chase*				
Mating: *Copulation*	M/F			
Nest-building	F			
Eggs and Young: *Tail flick*				
Eggs and Young: *Warning call*	M			
Eggs and Young: *Warning call*	F			
Eggs and Young: *Wing-flipping*	F			
Eggs and Young: *Eggs visible in nest*	N/A			
Eggs and Young: *Young visible in nest*	N/A			

Notes: _____

Red-Winged Blackbird Behavior Data Sheet

Observer Name _____

Date _____

Location _____

Habitat description _____

Weather _____

Total number birds observed: males _____ females _____

Observed Behavior (*Stokes 1979*)	Sex	Frequency of Occurrence	Duration (minutes)	Vegetation Type (woody vs. grassy)
Territory: *Songspread*				
Territory: *Song-flight*				
Territory: *Bill-Tilt*				
Courtship: *Crouch*	M			
Courtship: *Sexual-chase*				
Mating: *Copulation*	M/F			
Nest-building	F			
Eggs and Young: *Tail flick*				
Eggs and Young: *Warning call*	M			
Eggs and Young: *Warning call*	F			
Eggs and Young: *Wing-flipping*	F			
Eggs and Young: *Eggs visible in nest*	N/A			
Eggs and Young: *Young visible in nest*	N/A			

Notes: _____

# 13 A Field Investigation on the Seed Foraging Behavior of Vertebrates in the Natural Environment

Purpose

To be able to describe how variations in the natural environment may differentially affect the foraging behavior of wild animals.

To develop skills in quantitative analysis and critical thinking.

Materials

80 plastic Petri dishes (ca. 10-cm diameter)
Wire (ca. 10 gauge) that can be bent for staking down Petri dishes (2–3 per dish)
160 Rubber bands
Sunflower seeds (at least 100 g)
Fine dry sand (at least 4 L or 25 lbs.)
Permanent marker
Triple beam balances or electronic pan scale (with ± 0.01 g accuracy)
Graduated cylinder
Disposable gloves
Forceps
Window screening
OSHA approved N95 particulate disposable respirator
Goggles

Optional:
Millet seeds, thistle seeds, or alfalfa pellets (at least 100 g each)

Introduction

Various factors influence behavior of organisms in their natural environment, to include microhabitat variation, food availability, interspecific competition, predation risk, and physiological constraints (Yunger et al. 2002). Through experimental field investigations, it is possible to examine one or more variables at a time to reveal ecological factors that help shape animal behavior. The activity patterns and other behaviors of many animals are structured in part by the need to forage (procure food). Foraging behavior of vertebrate animals can be examined thorough resource use (i.e., food consumption) under different conditions. This can be measured, in turn, by "giving-up density" (GUD), which is the amount of resource remaining following use by an organism (Brown et al. 1997).

In this investigation, you will attempt to describe foraging patterns of small mammals in different habitats of a natural area. Common small mammals of grassland and woodland of the Midwest include the white-footed mouse (*Peromyscus leucopus*), the meadow mole (*Microtus pennsylvanicus*), prairie vole (*M. ochrogaster*), and eastern chipmunk (*Tamias striatus*). The white-footed mouse is abundant in both woodland and grassland, eastern chipmunks in woodland, and voles in grasslands. All are granivorous to some degree.

You must generate at least one hypothesis relevant to the investigation. Experimental sites are to be located in grassland or woodland. Restored habitats are suitable sites for study. Your instructor will provide a history of the experimental sites to include management (e.g., fire frequency and timing of burns).

Procedures

Part 1—Methods for assessing foraging behavior

1. Choose a partner with whom to work. Uniquely number the bottom of each Petri dish using permanent marker (numbers 1–80). Additionally label dishes 1–40 with "COV," for vegetation cover and dishes 41–80 "NOCOV," for little to no vegetation cover. Divide the 80 dishes equally among pairs of students.

2. Measure and place 50 ml of fine sand in the bottom of each Petri dish. The sand will represent non-nutritive substrate through which animals must sift to locate food. Measure 1.00 g of sunflower seeds and mix into sand in each dish. *Be sure both sand, sunflower seeds, and insides of Petri dishes are dry.* Rubberband together the lid and bottom of each dish (i.e., foraging tray). Weigh each entire dish and its contents. Record weights prior to transport to the field site.

3. Seed trays will be placed in different habitats during a time when precipitation is not forecasted for at least 24 hr. Locate areas that differ substantially in the amount of vegetation cover from a rodent's perspective (i.e., a lot to no or very little herbaceous and low woody vegetation). Place the foraging trays in their appropriate habitat (cover vs. no cover) by removing the lids and nesting them under the bottom. Space the foraging trays at least 3 m apart. Use 2–3 wire hooks to stake down each tray; you may also want to stake down the set of rubberbands, which you will need to transport trays from the field site back to the lab. Seed trays should remain in the environment overnight (ca. 12 hr) to allow nocturnal rodents to feed. *Note: If time permits, introduce sealed trays to the environment for several days prior to opening up for foraging. This will give rodents a chance to acclimate to the novel objects.*

   Suggestions for optional designs: (1) Use paired trays (side-by-side) to assess seed size preferences, by including relatively large seeds in one tray and relatively small seeds in the other tray. (2) Use paired trays to assess food type preferences, by including one type of seed in one tray and alfalfa pellets in the other tray. (3) Partition rodent vs. bird foraging on the basis of their different activity patterns. Small mammal seed predators such as mice forage generally at night (dusk to dawn), birds such as sparrows and finches, during the day (dawn to dusk).

4. Collect trays for transport back to the lab following your allotted foraging time in the field. Make note if sand was visibly lost from any trays and from which ones.

Important: some diseases can be transmitted to humans from rodent excrement. Avoid direct contact and inhalation of dust from foraging trays; use respirator, goggles, and disposable masks as a precaution when handling trays after exposure to wild rodents. Instructor should provide background information on potential diseases and appropriate methods to avoid transmission.

Wear protective gloves, an N95 particulate disposable mask or respirator, and goggles. Take care not to disturb and breathe in dust from contents of dishes. Seal the remaining contents of each dish with its lid and rubber bands for transport back to the lab.

Part 2—Method for obtaining giving up densities

1. When conducting the following procedure, you should wear goggles and a respirator, in addition to protective gloves, if there is any chance of inhalation. In an exhaust hood and wearing protective gloves, weigh each tray with its remaining contents. Be sure there is no moisture on the outside of the tray. If you initially weighed the tray with the rubber bands, leave rubber bands on during the final weighing. This final weight is the GUD for each tray. Exclude trays that had visible, significant sand loss. If the final weight is greater than the initial weight of seeds, the contents of the foraging trays have likely absorbed moisture. If so, proceed to Step 2.
2. Place trays under the exhaust hood or in a drying oven ventilated to the outside. Remove lids from the foraging trays and allow sand and remaining seeds to completely dry until no more moisture is retained, i.e., a constant weight for a particular tray is achieved.

Part 3—Data analysis

Compute the mean GUD and a measure of its variance, such as standard deviation, for each experimental variable. Use a t-test (see the lab "An Introduction to Biometry" in this manual) to compare these means and to ascertain if your results supported your hypothesis. You may either use a computer program or perform the calculations by hand. Report results of your calculations and conclusion. Your conclusion should offer an explanation of the results and provide critical comment as to data support of your initial hypothesis.

Suggested Activity

Write a scientific paper on this lab investigation. Be sure to follow the guidelines given by your instructor and those in the lab, "Writing a Scientific Paper."

References

Brown, J. S., B. P. Kotler, and W. A. Mitchell. 1997. Competition between birds and mammals: A comparison of giving-up densities between crested larks and gerbils. Evolutionary Ecology 11:757–771.

Yunger, J. A., P. L. Meserve, and J. R. Gutierrez. 2002. Small-mammal foraging behavior: mechanisms for coexistence and implication for population dynamics. Ecological Monographs 72:561–577.

EXERCISE
A Field Investigation on the Seed Foraging Behavior of Vertebrates in the Natural Environment

Name_____ Date_____

Questions

1. List and describe three different factors that can affect the amount of time a small mammal or bird spends foraging in a particular location.

2. Describe a scenario that could possibly contribute to high giving-up density (large amount of food remaining) for a mouse in a foraging patch.

3. State your hypothesis. Perform a t-test; show your work. State whether your results support your hypothesis or not.

Foraging Experiment Data

Tray No.	Experimental Variable	Initial Seed Wt. (g)	Final Seed Wt. (g) = GUD
1	COV	5.00	
2	COV	5.00	
3	COV	5.00	
4	COV	5.00	
5	COV	5.00	
6	COV	5.00	
7	COV	5.00	
8	COV	5.00	
9	COV	5.00	
10	COV	5.00	
11	COV	5.00	
12	COV	5.00	
13	COV	5.00	
14	COV	5.00	
15	COV	5.00	
16	COV	5.00	
17	COV	5.00	
18	COV	5.00	
19	COV	5.00	
20	COV	5.00	
21	COV	5.00	
22	COV	5.00	
23	COV	5.00	
24	COV	5.00	
25	COV	5.00	
26	COV	5.00	
27	COV	5.00	
28	COV	5.00	
29	COV	5.00	
30	COV	5.00	
31	COV	5.00	
32	COV	5.00	
33	COV	5.00	
34	COV	5.00	
35	COV	5.00	
36	COV	5.00	
37	COV	5.00	
38	COV	5.00	
39	COV	5.00	
40	COV	5.00	
Grand Total			
Mean			

Foraging Experiment Data

Tray No.	Experimental Variable	Initial Seed Wt. (g)	Final Seed Wt. (g) = GUD
41	NOCOV	5.00	
42	NOCOV	5.00	
43	NOCOV	5.00	
44	NOCOV	5.00	
45	NOCOV	5.00	
46	NOCOV	5.00	
47	NOCOV	5.00	
48	NOCOV	5.00	
49	NOCOV	5.00	
50	NOCOV	5.00	
51	NOCOV	5.00	
52	NOCOV	5.00	
53	NOCOV	5.00	
54	NOCOV	5.00	
55	NOCOV	5.00	
56	NOCOV	5.00	
57	NOCOV	5.00	
58	NOCOV	5.00	
59	NOCOV	5.00	
60	NOCOV	5.00	
61	NOCOV	5.00	
62	NOCOV	5.00	
63	NOCOV	5.00	
64	NOCOV	5.00	
65	NOCOV	5.00	
66	NOCOV	5.00	
67	NOCOV	5.00	
68	NOCOV	5.00	
69	NOCOV	5.00	
70	NOCOV	5.00	
71	NOCOV	5.00	
72	NOCOV	5.00	
73	NOCOV	5.00	
74	NOCOV	5.00	
75	NOCOV	5.00	
76	NOCOV	5.00	
77	NOCOV	5.00	
78	NOCOV	5.00	
79	NOCOV	5.00	
80	NOCOV	5.00	
Grand Total			
Mean			

14 Pollination and the Distribution of Plants

Purpose
To gain understanding of how pollination can affect the distribution of plants and to develop critical thinking skills.

Materials
Timing devices that can counts seconds (e.g., watches), calculator, and access to two populations of flowering plants that should exhibit clumped and sparse patterns of distribution in the same local area.

Introduction
Much of the speciation of the seed producing plants has been attributed to the symbiotic relationships that these plants have with insects. An example of a mutualistic symbiosis is pollination. Insects pollinate by transferring pollen to the stigma of the carpel. In return, insects receive nutrition from the flower in the forms of nectar, pollen and/or other tissues of the flower (e.g., petals or sepals). Nectar is a sugary liquid secreted from nectaries which are associated with phloem.

Many flowering plants compete for insect pollinators (Brody 1997, Hainsworth et al. 1984, Wright and Meagher 2003). This competition explains the differences in flowering times among angiosperms and the conspicuous flower displays in the forms of scents and color. It seems logical that for plants having overlapping flowering times or less conspicuous flowers, the ability to clump together would be selectively advantageous as the clump should better draw the attention of the pollinators to the flowers as compared to plants that are more sparsely distributed. Probability of pollination should increase with higher frequencies and longer durations of pollinator visits and, especially among plants having composite inflorescences or flowers having multiple carpels. Here you will test if clumping acts to encourage flower visitation by insect pollinators in one or more species of angiosperms.

Procedures
Break into the number of groups dictated by the number of plant species you are to examine. Each group is to choose a different plant species that is currently flowering. Each group is to then divide into two subgroups for a given plant species, where one subgroup is to observe individual plants that are visibly clumped in distribution while the other subgroup is to observe those which are sparsely distributed in the same local environment. Each subgroup is to record the following over a pre-determined observation period (e.g., 20 minutes).

1. The number of plants being observed.
2. The frequency and length, in seconds, of visits by insects over the observation period. If an insect touches a flower, flies away, and returns, count this as two visits.

From these measurements, compute the number of insect visits per plant and the mean length of a visit.

References

Brody, A. 1997. Effects of pollinators, herbivores, and seed predators on flowering phenology. Ecology 78:1624–1631.

Hainsworth, F. R., L. L. Wolf, and T. Mercier. 1984. Pollination and pre-dispersal seed predation: net effects on reproduction and inflorescence characteristics in *Ipomopsis aggregata*. Oecologia 63:405–409.

Wright, J. W. and T. R. Meagher. 2003. Pollination and seed predation drive flowering phenology in *Silene latifolia* (Caryophyllaceae). Ecology 84:2062–2073.

EXERCISE
Pollination and the Distribution of Plants

Name_____   Date_____

Plant Species	Type of Distribution	Sum of All Time Spent Visiting in Seconds	Number of Visits per Plant	Average Visit Length in Seconds
	Clumped			
	Sparsely distributed			
	Clumped			
	Sparsely distributed			
	Clumped			
	Sparsely distributed			

1. What are the advantages of clumped distribution in a population of angiosperms? Be sure to not limit your answer with respect to pollination ecology.

2. What are the advantages of sparse distribution in a population of angiosperms? Be sure to not limit your answer with respect to pollination ecology.

3. If one type of distribution by a particular angiosperm species appears to be advantageous, why are individuals distributed otherwise?

15 A Laboratory Investigation of Population Growth

Purpose

To assess how populations increase over time when individuals compete for limited resources.

To gain a practical understanding of these ecological concepts: exponential population growth, logistic population growth, carrying capacity (K), biotic potential (r), and intraspecific competition.

Materials

Lemna minor
filtered pond water
test tubes, 15 cm × 2 cm
grease pencils
plastic wrap and rubber bands
test tube rack
fluorescent light stand
light meter
class Data Sheet
scientific calculators
arithmetic graph paper

Introduction

Population ecologists have produced models to demonstrate how populations grow over time in favorable environments. Because of the many variables that could influence population growth, these models may not describe how all populations will grow, but are useful, general representations. If resources (e.g., food, water, space) were unlimited, a population might grow exponentially (Figure 15.1). We can use a differential equation to describe exponential growth:

$dN/dt = rN$ eqn. 1

or the integrated formula:

$N_t = N_0 e^{rt}$ eqn. 2

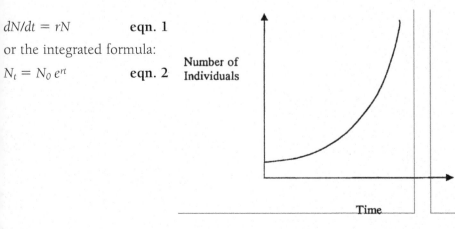

Figure 15.1
An exponential growth curve (J-shaped).

141

where

N_t = the number of individuals at time t
N_0 = the number of individuals at time 0, or the start of population monitoring
e = 2.71828
r = the intrinsic rate of natural increase (biotic potential; maximal per capita growth rate)
t = time

Since an unlimited population increase is not realistic in nature (Why?), we can introduce a new variable, K, the carrying capacity or the maximum amount of individuals of a species that the environment can sustain over time. The resulting formulae describe logistic population growth (Figure 15.2):

$$dN/dt = rN\,(K-N/K) \qquad \text{eqn. 3}$$

or:

$$N_t = K/1 + e^{a-rt} \qquad \text{eqn. 4}$$

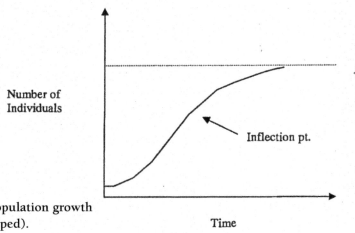

Figure 15.2
A logistic population growth curve (S-shaped).

where

a = a constant of integration defining the position of the curve relative to the origin (the y-intercept).

You will study the population growth of the duckweed, *Lemna minor*. Duckweeds are small, floating, vascular plants that grow in calm fresh waters, such as local ponds. Individual plants consist of small fronds (flat, leaf-like structures) with small visible roots. Fronds are often clustered. Duckweeds may flower but mainly reproduce asexually (vegetatively) by budding, a characteristic useful for our population growth investigation.

Procedure

Part 1—field and laboratory work

1. Obtain living duckweed from a pond. Collect some of the pond water.
2. Pass the pond water through a course filter in the laboratory. Save the filtered water for filling the test tubes in which you will grow the duckweed.

3. Each pair of students will prepare 1 test tube. Fill the test tube 2 cm from the top with filtered pond water. Add a cluster of 3 duckweeds. Cover the tube with the plastic wrap held in place by a rubber band, and poke 5 small holes in the wrap. Label your tube with your group number. Place the tube in a test tube rack under fluorescent lights. Adjust so that the plants receive 400 foot-candles of light (ca. 4,300 lumens/m^2). The lights will be on continuously and plants will be grown at room temperature or slightly warmer (ca. 27° C). Alternatively, plants may receive 800 foot-candles of light for a 12-hr cycle of light/dark.

4. Record starting number of plants on data sheet. You and your partner (you may alternate) will count and record the number of individuals in your test tube at least 3 times per week (5 if possible for the 1st week) at about the same time each day for 3–4 weeks. Place your initials next to your recorded value.

Part 2—data analysis

1. At the end of 3–4 weeks, pool the class data by taking the mean number in all test tubes for each day. <u>Record on your Summary Data Sheet</u>. The means will be used for your calculations and graphs.

2. <u>Plot the mean number of individual duckweeds vs. time</u> and notice if the curve looks like the exponential or logistic growth curve. Use either a computer program or graphing paper (instructor may provide in class).

3. To determine when growth rate was highest (inflection point on S-shaped curve), calculate the change in numbers for all test tubes between days, ΔN, and the change in numbers per day, $\Delta N/t$, (<u>place these values on your Summary Data Sheet</u>). Plot $\Delta N/t$ vs. time. Indicate on your graph during which daily interval growth rate was the highest.

4. Next you will estimate r, the biotic potential or maximal per capita growth rate for the duckweed. Check the daily interval with the highest growth rate (estimated in number 3 above) by calculating the per capita growth rate, $(\Delta N/t)/N$, for each day (<u>place on your Summary Data Sheet</u>). Use the highest daily interval (it should match number 3 above) to calculate an estimate of r. To do this, use the rearranged formula of eqn. 2:

<u>Report this estimate on your graph.</u>

5. Once your growth curve levels off, use the number of plants for K, the carrying capacity in your test tube. Now you will try to <u>fit</u> the data to the logistic model $N_t = K/1 + e^{a-rt}$ (eqn. 4) by <u>calculating and plotting new N_t values</u>. You already have r and K, but need a so you know where to start the curve on your graph. Rearrange eqn. 4 to solve for a:

eqn. 5

where $t = 0$.

Calculate enough N_t values (ca. every 2–3 time periods) to generate a logistic curve. Record these values <u>on your Summary Data Sheet</u> for the days in which you calculated N_t for the logistic fit. <u>Plot this curve on the same graph as your original data</u>, N_t vs. time, and compare.

Suggested Activity

Consider adding another variable to the experiment such as the addition of fertilizer or the replacement of pond water with distilled water.

Write a scientific paper on this lab investigation. Be sure to follow the guidelines given by your instructor and those in the lab, "Writing a Scientific Paper."

Adopted from:

Brewer, R. and M. T. McCann. 1982. Laboratory and Field Manual of Ecology. Saunders College Publishing, Philadelphia, PA, USA.

EXERCISE
A Laboratory Investigation of Population Growth

Name_____ Date_____

Questions

1. What is an overshoot of carrying capacity? Did your population exhibit this? Justify your answer by comparing K for your actual growth curve to the logistic curve you generated.

2. Briefly discuss whether you found the duckweed counting method to be effective. What other suggestions on counting methods do you have to improve this experiment?

3. What types of organisms might exhibit exponential growth in nature (to a point)?

4. Show your work for your estimate r, the biotic potential or maximal per capita growth rate for the duckweed. Use the formula from (4) in the Data Analysis section.

5. Does the r value you estimated (in 4 above) appear too high or low?

6. Discuss life history factors of duckweed (how it grows and reproduces in nature) and biotic potential. Compared to other organisms, is the biotic potential of duckweed relatively high or low?

7. What are advantages/disadvantages to having a high r?

8. Extra: Can you think of a better way to calculate r such as using all of your data? (Hint: rearrange eqn. 5 to get the equation for a straight line where r is the slope.)

Lemna Population Growth—Class Data Sheet

Name/Group →													
Day 1													
Day 2													
Day 3													
Day 4													
Day 5													
Day 6													
Day 7													
Day 8													
Day 9													
Day 10													
Day 11													
Day 12													
Day 13													
Day 14													
Day 15													
Day 16													
Day 17													
Day 18													
Day 19													
Day 20													
Day 21													
Day 22													
Day 23													
Day 24													
Day 25													
Day 26													
Day 27													
Day 28													
Day 29													
Day 30													

Lemna Population Growth—Summary Data Sheet

Name_____   Class Section_____

	No. in Your Test Tube, N	Mean No. for All Tubes, N	ΔN, Between Days, for All Tubes	Change in No. per Day $\Delta N/t$	Per Capita Growth Rate, $(\Delta N/t)/N$	Logistic Fit of $N_t =$ $K/1 + e^{a-rt}$
Day 1						
Day 2						
Day 3						
Day 4						
Day 5						
Day 6						
Day 7						
Day 8						
Day 9						
Day 10						
Day 11						
Day 12						
Day 13						
Day 14						
Day 15						
Day 16						
Day 17						
Day 18						
Day 19						
Day 20						
Day 21						
Day 22						
Day 23						
Day 24						
Day 25						
Day 26						
Day 27						
Day 28						
Day 29						
Day 30						

16 Life Tables

Purpose

To learn how life tables are constructed and to use information from life tables to gain insight into the dynamics of populations.

Materials

Calculator

Introduction

Life tables summarize mortality and survivorship data for populations. This information can be used to develop management strategies for wildlife and pests, or by insurance companies to establish premiums. In the following exercise, a cohort life table is to be developed. A cohort life table follows a group of animals born within a discrete period of time to the death of the last individual. The demographics for the study populations were taken from Campbell (1969) and Lowe (1969).

Parameters of the cohort life table and their quantification are below.

N_x = the number of individuals in a cohort alive at age x.
d_x = the number of individuals in a cohort that die in the x age interval
 = $(N_x - N_{x+1})$.
q_x = age specific mortality rate = d_x/number alive at beginning of the time interval.
L_x = the average number of surviving individuals during the time interval = $(N_{x+1} + N_x)/2$.

After computing the data above, evaluate the remaining parameters.

T_x = sum L_x column cumulatively from the bottom up.
e_x = future life expectancy = T_x/N_x.

The variety of developmental stages shown by insects, like the gypsy moth (*Lymantria dispar*), makes for an interesting life table study. Introduced to North America in 1869 by a scientist researching silk production, the moth escaped confinement and over the past century, spread across the continent. Larvae defoliate and weaken trees of native forests and urban areas, causing millions of dollars of damage each year. The gypsy moth has also been found to be part of an intricate food web that includes the bacterium responsible for Lyme disease (Jones et al. 1998).

Table 16.1
Life Table for Gypsy Moths as Measured from a Sample Population in Northeastern Connecticut
The year is the unit of time.

Age (x)	N_x	d_x	q_x	L_x	T_x	e_x
Eggs	550	165	0.30	467.5	931.0	1.69
Larval Instars 1–3	385	142	0.37	314	463.5	1.20
Larval Instars 4–5	243	228	0.94	129	149.5	0.62
Prepupae	15	3	0.20	13.5	20.5	1.37
Pupae	12	11	0.92	6.5	7.0	0.58
Adults	1	1	1	0.5	0.5	0.5

Table 16.1 provides an example of computations using data for gypsy moths from Campbell (1969). Out of the 550 eggs oviposited, 385 were counted within the larval instar 1–3 cohort, indicating a mortality of (550–385) or 165 moths. The age specific mortality rate is computed as 165/550 or 0.30. The average number of surviving individuals during the time period of the first cohort was [(550 + 385)/2] or 467.5. Determine T_x from the oldest surviving cohort and proceed to the younger cohorts observed earlier in time. Beginning with the adult gypsy moth, T_x is computed as the L_x of the adult stage, or 0.5. The T_x for the pupae is then computed as the sum of T_x for the adult stage and L_x for the pupae (0.5 + 6.5), or 7.0. The T_x for the prepupae is computed as the sum of T_x for the pupae and the L_x for the prepupae (7 + 13.5), or 20.5. Calculate the life expectancy, e_x, as the quotient of T_x and N_x. The life expectancy, e_x, for the eggs is computed as 931/550, or 1.69 years.

Procedures

Complete the life tables on the data sheet for each gender of red deer (*Cervus elaphus*). The initial data were taken from Lowe (1969) who followed a cohort of the deer on the isolated Isle of Rhum beginning in 1957.

References

Campbell, R. W. 1969. Studies on gypsy moth populations dynamics. Page 29–34 *in* Forest Insect Population Dynamics. United States Department of Agriculture Research Paper, NE-125.

Jones, C.G., R.S. Ostfeld, M.P. Richard, E.M. Schauber, and J.O. Wolff. 1998. Chain Reactions linking acorns to gypsy moth outbreaks and Lyme disease. Science 279:1023–1026.

Lowe, V.P.Q. 1969. Population dynamics of red deer (*Cervus elaphus* L.) on Rhum. Journal of Animal Ecology 38:425–457.

EXERCISE
Life Tables

Name_____ Date_____

1. What can explain why the life expectancy of the prepupae cohort of gypsy moths (as shown in Table 16.1) is longer than that for either larval cohort?

2. How can life table analysis be used to manage gypsy moths?

Life Table for Red Deer Stags (males)
The year is the unit of time.

Age (x)	N_x	d_x	q_x	L_x	T_x	e_x
1	1000					
2	916					
3	897					
4	897					
5	747					
6	426					
7	208					
8	150					
9	20					

Life Table for Red Deer Hinds (females)
The year is the unit of time.

Age (x)	N_x	d_x	q_x	L_x	T_x	e_x
1	1000					
2	1000					
3	939					
4	754					
5	505					
6	305					
7	186					
8	132					
9	25					

3. The life expectancies of red deer stags and hinds as measured for the first year are less than the maximum life spans observed. Explain these discrepancies.

4. During what ages do red deer stags show relatively little changes in life expectancy (where the additional living of a year does not correspond to a major decrease in life expectancy)? What can explain this apparent "prolonging" of life expectancy in red deer stags?

Appendices

Appendix 1
Critical *t* Values at Selected Probabilities
df = degrees of freedom

df 2-Tailed: 1-Tailed:	0.10 0.05	0.05 0.025
1	6.314	12.706
2	2.920	4.303
3	2.353	3.182
4	2.132	2.776
5	2.015	2.571
6	1.943	2.447
7	1.895	2.365
8	1.860	2.306
9	1.833	2.262
10	1.812	2.228
11	1.796	2.201
12	1.782	2.179
13	1.771	2.160
14	1.761	2.145
15	1.753	2.131
16	1.746	2.120
17	1.740	2.110
18	1.734	2.101
19	1.729	2.093
20	1.725	2.086
21	1.721	2.080
22	1.717	2.074
23	1.714	2.069
24	1.711	2.064
25	1.708	2.060
30	1.697	2.042
40	1.684	2.021
50	1.676	2.009
60	1.671	2.000
70	1.667	1.994
80	1.664	1.990
90	1.662	1.987
100	1.660	1.984
∞	1.645	1.960

Appendix 2
Critical F Values for a Selected Probability of 0.05
($\alpha = 0.05$)

df of Denominator	df of Numerator													
	1	2	3	4	5	6	7	8	9	10	20	30	50	∞
1	161	200	216	225	230	234	237	239	241	242	248	250	252	254
2	18.51	19.00	19.16	19.25	19.30	19.33	19.36	19.37	19.38	19.39	19.44	19.46	19.47	19.50
3	10.13	9.55	9.28	9.12	9.01	8.94	8.88	8.84	8.81	8.78	8.66	8.62	8.58	8.53
4	7.71	6.94	6.59	6.39	6.26	6.16	6.09	6.04	6.00	5.96	5.80	5.74	5.70	5.63
5	6.61	5.59	5.41	5.19	5.05	4.95	4.88	4.82	4.78	4.74	4.56	4.50	4.44	4.36
6	5.99	5.14	4.76	4.53	4.39	4.28	4.21	4.15	4.10	4.06	3.87	3.81	3.75	3.67
7	5.59	4.74	4.35	4.12	3.97	3.87	3.79	3.73	3.68	3.63	3.44	3.38	3.32	3.23
8	5.32	4.46	4.07	3.84	3.69	3.58	3.50	3.44	3.39	3.34	3.15	3.08	3.03	2.93
9	5.12	4.26	3.86	3.63	3.48	3.37	3.29	3.23	3.18	3.13	2.93	2.86	2.80	2.71
10	4.96	4.10	3.71	3.48	3.33	3.22	3.14	3.07	3.02	2.97	2.77	2.70	2.64	2.54
20	4.35	3.49	3.10	2.87	2.71	2.60	2.52	2.45	2.40	2.35	2.12	2.04	1.96	1.84
30	4.17	3.32	2.92	2.69	2.53	2.42	2.34	2.27	2.21	2.16	1.93	1.84	1.76	1.62
50	4.03	3.18	2.79	2.56	2.40	2.29	2.20	2.13	2.07	2.02	1.78	1.69	1.60	1.44
∞	3.84	2.99	2.60	2.37	2.21	2.09	2.01	1.94	1.88	1.83	1.57	1.46	1.52	1.00

Appendix 3
Mann-Whitney Critical Values for a Selected Probability of 0.05
($\alpha = 0.05$ and two-tailed test)

n_2	n_1													
	2	3	4	5	6	7	8	9	10	11	12	13	14	15
2							0	0	0	1	1	1	1	1
3				0	1	1	2	2	3	3	4	4	5	6
4			0	1	2	3	4	4	5	6	7	8	9	10
5		0	1	2	3	5	6	7	8	9	11	12	13	14
6		1	2	3	5	6	8	10	11	13	14	16	17	19
7		1	3	5	6	8	10	12	14	16	18	20	22	24
8	0	2	4	6	8	10	13	15	17	19	22	24	26	29
9	0	2	4	7	10	12	15	17	20	23	26	28	31	34
10	0	3	5	8	11	14	17	20	23	26	29	33	36	39
11	0	3	6	9	13	16	19	23	26	30	33	37	40	44
12	1	4	7	11	14	18	22	26	29	33	37	41	45	49
13	1	4	8	12	16	20	24	28	33	37	41	45	50	54
14	1	5	9	13	17	22	26	31	36	40	45	50	55	59
15	1	5	10	14	19	24	29	34	39	44	49	54	59	64

Appendix 4
Chi-square Critical Values for a Probability of 0.05
($\alpha = 0.05$)

df	Chi-Square Critical Value
1	3.841
2	5.991
3	7.815
4	9.488
5	11.070
6	12.592
7	14.067
8	15.507
9	16.919
10	18.307
11	19.675
12	21.026
13	22.362
14	23.685
15	24.996
16	26.296
17	27.587
18	28.869
19	30.144
20	31.410
30	43.773
40	55.758
50	67.505